Topics in Applied Physics Volume 76

Springer
Berlin
Heidelberg
New York
Barcelona
Hong Kong
London
Milan
Paris
Singapore
Tokyo

Topics in Applied Physics Founded by Helmut K. V. Lotsch

Volume 1–48 are listed at the end of the book

Light Scattering in Solids VIII

Fullerenes, Semiconductor Surfaces, Coherent Phonons

Edited by M. Cardona and G. Güntherodt

With Contributions by
M. Cardona, G. C. Cho, T. Dekorsy, N. Esser,
G. Güntherodt, H. Kurz, J. Menéndez,
J. B. Page, W. Richter

With 86 Figures and 12 Tables

Springer

Professor Dr., Dres. h. c. Manuel Cardona
Max-Planck-Institut für Festkörperphysik
Heisenbergstr. 1
D-70569 Stuttgart, Germany

Professor Dr. Gernot Güntherodt
2. Physikalisches Institut
Rheinisch-Westfälische Technische Hochschule Aachen
Templergraben 55
D-52074 Aachen, Germany

ISSN 0303-4216
ISBN 3-540-66085-2 Springer-Verlag Berlin Heidelberg New York

Library of Congress Cataloging-in-Publication Data applied for.

Die Deutsche Bibliothek - CIP-Einheitsaufnahme
Light scattering in solids. - Berlin; Heidelberg; New York; Barcelona; Hong Kong; London; Milan; Paris; Tokyo: Springer

8. Fullerenes, semiconductor surfaces, coherent phonons. - 2000
(Topics in applied physics; Vol. 76)
ISBN 3-540-66085-2

Cover concept: Studio Calamar Steinen
Cover production: *design & production* GmbH, Heidelberg
Typesetting: Data conversion by Steingraeber Satztechnik GmbH, Heidelberg

SPIN: 10732772 57/3144/mf - 5 4 3 2 1 0 – Printed on acid-free paper

Preface

This volume is the eighth of a series (Topics in Applied Physics, Vols. 8, 50, 51, 54, 66, 68, 75, 76) devoted to inelastic light scattering by solids, both as a physical effect and as a spectroscopic technique. It appeared shortly after the publication of *Light Scattering in Solids VII* and can be considered to be its continuation.

Light Scattering in Solids VI (LSS VI) appeared in 1991, four years after the discovery of high-temperature superconductivity. By the time it appeared, inelastic (Raman) light scattering had established itself as one of the most powerful techniques for the investigation of electronic excitations, magnons, phonons, and electron–phonon interaction in the new high-temperature superconductors. Correspondingly, a chapter of LSS VI was devoted to Raman scattering in high-temperature superconductors. In the past eight years, and with the discovery of new families of high-T_c superconductors, Raman spectroscopy has continued to demonstrate its usefulness for the investigation and characterization of this class of materials. Exciting new materials, such as fullerenes and carbon nanotubes, porous silicon, and the colossal magnetoresistance manganites, as well as low dimensional spin systems exhibiting spin-Peierls transitions and spin gaps, have also shown themselves to be excellent candidates for the investigation by means of inelastic light-scattering spectroscopy. Progress in instrumentation has extended the capabilites of Raman spectroscopy in the directions of spatial microsampling and time-resolved spectroscopy. Increasing commercial availability of laser-based equipment producing subpicosecond pulses has led to the technique of "coherent phonons" which can be considered equivalent to conventional spontaneous Raman scattering but in the time domain instead of the frequency domain.

Chapter 1 of this volume contains an introduction with a review of the work described in previous volumes, a summary of the contents of the present volume, and a survey of some of the progress in other aspects of Raman spectroscopy, in particular in the field of semiconductor nanostructures (including the fractional quantum Hall effect, of Nobel fame), and in Raman spectroscopy of isotopically modified crystals. Chapter 2 is devoted to fullerenes, Chap. 3 to Raman spectroscopy of semiconductors, surfaces, and interfaces, and Chap. 4 to coherent phonons.

The authors would like once again to thank Sabine Birtel for secretarial help and skillful use of modern word processing techniques. Thanks are also due to the staff of Springer-Verlag, in particular Ms Friedhilde Meyer and Dr. Werner Skolaut for their unbureaucratic and skillful production of this volume. Last but not least, we would like to recall that this book will appear nearly 70 years after the discovery of the Raman effect in Calcutta (India), in 1928, and the award of the Nobel prize to Sir Chandrasekhara V. Raman in 1930. We would like to dedicate it to Sir Chandrasekhara's memory.

Stuttgart and Aachen,
September 1999

Manuel Cardona
Gernot Güntherodt

Contents

Contributors

Manuel Cardona
MPI für Festkörperforschung
Heisenbergstr. 1
D-70569 Stuttgart
Germany
e-mail: cardona@cardix.mpi-stuttgart.mpg.de

Gernot Güntherodt
RWTH Aachen
2. Physikalisches Institut
D-52056 Aachen
Germany
e-mail: gernot-guentherodt@physik.rwth-aachen.de

José Menéndez
Department of Physics and Astronomy
Arizona State University
Tempe, AZ 85287-1504
USA
e-mail: jose.menendez@asu.edu

John B. Page
Department of Physics and Astronomy
Arizona State University
Tempe, AZ 85287-1504
USA
e-mail: john.page@asu.edu

Norbert Esser
Technische Universität Berlin
Institut für Festkörperphysik
Sekretariat PN 6-1
Hardenbergstr. 36
D-10623 Berlin
Germany
e-mail: norbes@gift.physik.tu-berlin.de

Wolfgang Richter
Technische Universität Berlin
Institut für Festkörperphysik
Sekretariat PN 6-1
Hardenbergstr. 36
D-10623 Berlin
Germany
e-mail: rich0431@mailszrz.zrz.tu-berlin.de

Thomas Dekorsy
Institut für Halbleitertechnik
Rheinisch-Westfälische Technische Hochschule
Sommerfeldstr. 24
D-52056 Aachen
Germany
e-mail: dekorsy@iht-ii.rwth-aachen.de

Gyu Cheon Cho
IMRA America Inc.
1044 Woodridge Ave.
Ann Arbor, MI 48105
USA
e-mail: gcho@imra.com

Heinrich Kurz
Institut für Halbleitertechnik
Rheinisch-Westfälische Technische Hochschule
Sommerfeldstr. 24
D-52056 Aachen
Germany
e-mail: kurz@amica.rwth-aachen.de

1 Introduction

M. Cardona and G. Güntherodt

With 10 Figures

> The size of a cell in the lattice being small compared with the wave-length of the incident light, the crystal may for practical purposes be regarded as a continuous homogeneous medium of *uniform optical density* and can accordingly scatter no light. As thermal movement disturbs the uniformity of the medium and introduces local fluctuations of optical density, the medium is no longer homogeneous but shows irregular variations of refractive index, which though small, nevertheless in the aggregate, result in an appreciable scattering of the light traversing the medium. The intensity of this scattering can be calculated if the average magnitude of fluctuation of optical density is known.
>
> *C.V. Raman, in Molecular Diffraction of Light (University of Calcutta Press, 1922); reprinted in the Scientific Papers of Sir C.V. Raman (Indian Academy of Science, Bangalore, 1978) p. 122*

This is the eighth volume of the series *Light Scattering in Solids* which appears in the Springer collection *Topics in Applied Physics* (TAP; numbers of previous volumes: 8, 50, 51, 54, 66, 68, 75). Since volume VII [1.1] preceded the present one only by a few months, this Introduction will forego a review of the previous volumes. It can be found in [1.1]. We shall instead review the contents of this volume together with related important developments that have taken place since the appearance of *Light Scattering in Solids VI* [1.2]. Developments related to the articles in [1.1], covering light scattering by crystal field transitions of electrons in partially filled d- and f-shells [1.3] and by spin wave excitations in layered magnetic structures [1.4] can also be found in Chap. 1 of [1.1].

The quotation from the extensive work of Sir Chandrasekhara Venkata Raman given above [1.5] was written in 1922, well before the discovery, in 1928, of the effect that bears his name. It applies, of course, to Rayleigh (and possibly Brillouin) scattering but it should convey a feeling for the beauty and precision of Raman's scientific writing. At the same time, it suggests that he was, already then, rather close to the discovery of the Raman effect:

Topics in Applied Physics, Vol. 76
Light Scattering in Solids VIII
Eds.: M. Cardona, G. Güntherodt

had he considered instead of fluctuations of the macroscopic strain, fluctuations of internal unit cell coordinates (or molecular structure parameters) he would have, already then, *predicted* the inelastic scattering of light by optical phonons and molecular vibrations. He left this piece of his glory to the Austrian physicist A. Smekal [1.6] who worked out the theory of the Raman effect, on the basis of a two-level system, only one year after Raman wrote the text reproduced above. Renewed interest in the inelastic scattering of light, triggered by the award of the Nobel prize to A.H. Compton in 1927 [1.1], led to Raman's observation of light scattering by molecular vibrations in 1928 [1.7]. Raman's first report of the inelastic scattering of light [1.7] was followed within a few weeks by an independent report of a similar discovery by two Russian scientists [1.8]. For his discovery, the Nobel prize in physics was awarded to C.V. Raman, G.S. Landsberg and L.I. Mandelstam in 1930, thus being, within two years after discovery, one of the fastest awards of such important recognition. This decision of the Nobel committee appears, in retrospect, to have been very appropriate. Raman spectroscopy has developed, especially since the advent of the laser in the 1960s, into one of the most powerful experimental techniques available not only to physicists and chemists but also to engineers, geologists, biologists, and to the medical profession. For a glimpse into the wide range of applicability of Raman spectroscopy the reader should consult the proceedings of the International Conferences on Raman Spectroscopy (ICORS), in particular the most recent one held in Cape Town (South Africa) in 1998 [1.9,1.10]. It is also of interest to note that Raman's Nobel prize is the only such distinction in physics awarded for work performed in a developing country, actually a country under colonial rule.

1.1 Contents of the Present Volume

1.1.1 Chapter 2

Chapter 2 of this volume is concerned with the vibrations of C_{60} molecules, the so-called Buckminsterfullerenes[1] or fullerenes for short. The internal molecular vibrations of these molecules, as well as the intermolecular modes of C_{60} crystals, are considered with emphasis in the former. Experimental data, obtained not only by means of Raman spectroscopy but also with ir-spectroscopy, inelastic neutron scattering, and electron energy loss spectroscopy, are presented. The observed vibrational frequencies, and the intensities of the corresponding ir as well as Raman modes, are discussed on the basis of group theory as well as various lattice dynamical models and *ab initio* electronic structure calculations. Isotope effects resulting from the isotopic abundances of natural carbon ($^{12}C_{0.99}{}^{13}C_{0.01}$) and also for samples synthesized with different isotopic compositions, are shown to be very useful for the assignment and interpretation of the observed modes.

[1] Named after the American architect R. Buckminster Fuller [1.11] who often used C_{60}-like shapes in his constructions.

The discovery of icosahedral C_{60} molecules resulted from a close collaboration between astronomers and molecular spectroscopists in an effort to understand unusual ir spectra found in carbon clusters emitted by red giant carbon stars [1.12, 1.13]. The experimental work of Kroto et al. [2.1] firmly established the existence in the laboratory of C_{60} molecules with icosahedral structure (I_h point group). These authors were able to synthesize the C_{60} molecules by means of laser ablation of a graphite target[2]. A few years later, Krätschmer et al. [2.2] were able to synthesize large amounts of C_{60} and make single crystals by evaporation of the solvent from a C_{60} solution in benzene [1.14].

With the work of Krätschmer et al. in 1990, C_{60} powders, films and crystals became generally available and an avalanche of research in the field was set in motion. A considerable part of this work was concerned with the investigation of vibrational properties, mostly by means of optical spectroscopies. The C_{60} molecule (possessing I_h point group symmetry) has a center of inversion and 46 different vibrational frequencies. Of these frequencies 10 correspond to Raman-active modes while only four are ir active. The remaining modes are silent and thus must be observed in higher-order Raman or ir spectra or with other types of spectroscopies. Table 2.4 of Chap. 2 gives recommended values for the frequencies of all intramolecular modes of C_{60}, 29 among them obtained rather reliably from experiments while the recommended values of the other 17 rely heavily on theoretical calculations.

Besides the vibrational frequencies of the C_{60} molecules, Chap. 2 is also concerned with the relative and also the absolute scattering cross sections, from both the experimental and the theoretical point of view. For this purpose one must distinguish between non-resonant and resonant cross sections. The most pronounced resonances occur for laser frequencies around 2.3 eV (see Fig. 2.14). They are likely to correspond to electronic transitions between the Highest Occupied and the Lowest Unoccupied Molecular Orbitals (HOMO and LUMO) although the details are not yet fully understood.

The interest in the C_{60} molecules and their derivatives has been greatly boosted by the discovery of superconductivity (having T_c up to nearly 40 K for samples placed under pressure) in C_{60} crystals doped with alkali metals (e.g., Rb_3C_{60}). Like in the case of high-T_c superconductivity, electronic Raman scattering has been recently shown to yield information about the superconducting (pair breaking) gap 2Δ [1.15]. Since this topic has not been covered in Chap. 2 of this volume, we show in Fig. 1.1 the spectra of Rb_3C_{60} at three temperatures, two below $T_c = 31$ K and one above. In order to eliminate sharp peaks due to phonons, Fig. 1.1 also displays the ratio of the scattering intensities below ($T = 4.5$ K) and above ($T = 35$ K) T_c. The step

[2] For their work on C_{60} curl, Kroto and Smalley were awarded the 1996 Nobel prize in chemistry. Also in 1996, but prior to receiving the Nobel prize, Sir Harald Kroto had been raised to a knighthood by Queen Elizabeth II. Before him, Sir Chandrasekhara V. Raman had been similarly honored.

in this ratio observed at $85 \pm 10\ \mathrm{cm}^{-1}$ has been assigned to 2Δ by the authors of [1.15]. This assignment leads to a value of $2\Delta/kT_c = 4.0 \pm 0.5$, compatible with that predicted by the BCS theory ($2\Delta/kT_c = 3.5$).

After the initial discovery of C_{60} and the methods to prepare it in macroscopic amounts, a number of other related molecules and materials were prepared. The most pervasive among them is C_{70}, an elongated "bucky ball" with the lower point group D_{5h}. Because of the lower symmetry, the vibrational spectra of C_{70} are considerably more complicated than those of C_{60}. They are treated briefly in the chapter under discussion. A large family of "materials" related to C_{60} and C_{70} is constituted by the so-called carbon nanotubes [2.11]. They are formed with rolled up graphitic sheets. Although they are not treated in the present work, Raman scattering also plays an important role in ongoing investigations of these nanotubes. The reader will find some information on the Raman scattering of nanotubes in Sect. 19.7 of [2.11]. For recent work on the polarized Raman spectra of single-wall carbon nanotubes see [1.16].

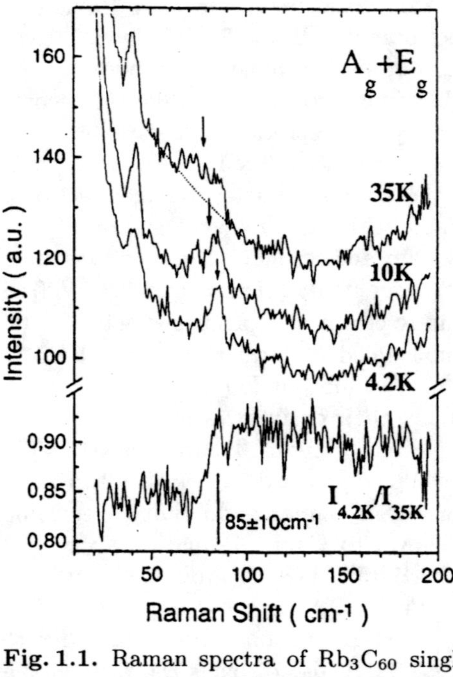

Fig. 1.1. Raman spectra of Rb_3C_{60} single crystals having $T_c = 31$ K at three temperatures: 35 K (above T_c), 10 K and 4.2 K (both below T_c). The lowest curve, labeled $I_{4.2K}/I_{35K}$, represents the ratio of the scattered intensity at 4.2 K to that at 35 K. The measurements were performed in the $A_g + E_g$ scattering configuration of the O_h point group of the crystal. The arrows indicate the estimated position of a T_{2g} phonon. The vertical bar at the bottom indicates the estimated position of the gap 2Δ [1.15]

1.1.2 Chapter 3

Chapter 3 of this volume, by Norbert Esser and Wolfgang Richter, discusses the application of Raman spectroscopy to the investigation of the vibrations of clean surfaces and monolayers of atoms deposited on them. The presentation gravitates around low-index surfaces ([100], [110], [110]) of III–V semiconducting compounds and of silicon. Heterojunctions are also treated.

The spectra discussed in Chap. 3 were taken, as is often the case, with visible lasers. For the semiconductors under consideration (e.g., Si, InAs, InP, ...) the penetration depth of visible light lies between 50 and 1000 nm, corresponding to 20 to 500 atomic monolayers. It was believed until recently that, under these conditions, only light scattering by bulk-like excitations should be observed. The first couple of monolayers at a crystal surface, however, have vibrational properties quite different from those of the bulk material since the restoring force constants should also be different. The restoring forces corresponding to both sides of the first layer should be rather different: the outer side has, naively speaking, no restoring forces. Moreover, surface relaxation and reconstruction should also introduce differences in force constants.

These changes in force constants lead to vibrational modes localized near the surfaces, that become *resonances* if they happen to be degenerate in energy with bulk modes. A crystal surface presents some form of two-dimensional translational lattice, being invariant upon a translation by a vector of this lattice (lying in the plane of the surface under consideration). In the third direction of space, perpendicular to the surface, the translational symmetry is broken. Consequently, the surface vibrational modes can be classified according to a surface $\boldsymbol{k}$-vector ($\boldsymbol{k}_{\perp}$) and the corresponding two-dimensional Bloch theorem. For computational purposes, 3-dimensional translational symmetry is often restored by considering a one-dimensional array of two-dimensional layers. This array must be constructed in such a way that the layers, separated by vacuum, are sufficiently thick and separated from their neighbors so that no significant interaction between the surface modes occurs. This technique, pioneered by *de Wette* [3.16], can be applied to either semiempirical lattice dynamical models or *ab initio* calculations that start from the full electronic band structure (see Sect. 3.2).

The type of surface modes we had in mind in the above discussion, resulting from short-range force constants, are the so-called microscopic surface modes. These modes, having typically optical phonon character (atoms vibrating *against* their neigbors) are confined to only very few (two to four) layers near the surface. Besides these microscopic surface phonons, other types exist that can be obtained from the solution of macroscopic equations in which the material properties are represented by macroscopic constants and/or dielectric functions. The prototype of macroscopic modes based on the elastic constants are the so-called Rayleigh modes. They are found in the frequency region of long-wavelength acoustic modes and appear in polar as well as in non-polar materials (see Fig. 7.2 of [1.17] and Fig. 6.14 of [1.18]).

The modes based on the dielectric function are found only for polar materials, in the region between the TO and LO frequencies of an ir-active phonon [3.24]. These modes are obtained when the appropriate boundary conditions are applied to the half space occupied by the crystal represented by the differental equations of elasticity and by Maxwell's equations [1.19]. They also correspond to Bloch-wave propagation along the surface and are characterized by a penetration depth into the crystal of the order of the in-plane surface wavelength (see (3.39) of [1.17]). Therefore the penetration depth of these waves can be very large (hence the designation *macroscopic*), a fact that makes them observable by optical spectroscopies. Macroscopic waves of the same nature are also found at the interfaces of heterojunctions and superlattices. These *interface modes* have been discussed in [1.19, 1.20]. Examples of their dispersion relations are shown in Figs. 1.5, 6 of [1.2]. Recent developments will be presented in Sect. 2.2 of this chapter.

Chapter 3 of this book is mainly concerned with microscopic surface vibrations (two-dimensional phonons) which, as already mentioned, are confined to a few atomic layers at the surface of a semiconductor. Because of their short penetration depth compared to that of the light of a visible laser, it was generally thought till recently that vibrations of only a couple of monolayers are not accessible to Raman and ir spectroscopy. Techniques such as High-Resolution Electron Energy-Loss (HREELS) [3.33–35] and Helium-Atom Scattering (HAS) (HAS) [3.20, 36, 37], with a penetration depth of the order of one monolayer, are ideal for the investigation of surface vibrations. Nevertheless, their resolution is limited and the required experimental equipment and know how is rather extensive. Therefore, several workers, among them *Chabal* [3.38] in the case of ir-spectroscopy and the group of *Richter* for Raman spectroscopy, have made efforts to enhance the sensitivity of optical techniques in order to allow the observation of surface-localized modes. In the case of Raman spectroscopy, considerable enhancement of the scattering cross section is possible when either the laser or the scattered frequency (or both) is close to that of strong optically active electronic interband transitions, resulting in the so-called resonant Raman effect. This enhancement is particularly useful when surface electronic transitions are well separated in frequency from bulk transitions and correspondingly, a resonance can be chosen to enhance mainly the scattering by surface vibrations. Moreover, when the frequency of these vibrations is different from that of bulk modes, an additional capability to discriminate surface from bulk vibrations appears. The separation of surface from bulk vibrations is particularly strong when the vibration of adsorbed monolayers of atoms with masses rather different from those of the bulk are being investigated. Chapter 3 of this book discusses the results of this type of work involving surface monolayers of Sb adsorbed on various III–V semiconductor surfaces (e.g., InP, GaAs). Not only the vibrational frequencies but also the corresponding scattering efficiencies (i.e., cross sections) are considered and compared with recent theoretical predictions. In

Sect. 3.4.1 the surface vibrations of arsenic-terminated silicon are discussed and corresponding resonant Raman scattering (RRS) data are presented. Section 3.4.2 discusses Raman data obtained for hydrogen-terminated (111) silicon surfaces. By saturating dangling bonds such hydrogen *passivates* the electronic properties of the (111) Si surfaces. Slight deviations from the (111) orientation result in (111) terraces and perpendicular step edges which can accommodate four different Si–H bonding configurations. In this case, four peaks are seen in the Raman spectrum: one corresponding to the stretching of Si–H bonds located on terraces (at $\approx$2080 cm^{-1}) and three corresponding to the step edges (C_1, C_2, C_3 in Fig. 3.25). It is of interest that these vibrations are seen in the Raman spectra, although no resonance attributable to the Si–H bonds occurs in the visible region (they take place in the uv). Nevertheless, the scattering efficiencies of the Si–H on bulk silicon are much higher than those of the same bonds in free molecules (e.g., SiH_4) thus indicating large coupling of the Si–H stretching vibrations to the electronic states of bulk silicon.

After describing the sulfur-related vibrations of a sulfur covered (100) surface in Sect. 3.4.3, the authors of Chap. 3 turn their attention, in Sect. 3.5, to the detection of the *surface phonons* of a clean (110) surface of InP. Such a surface shows, in all III–V compounds, a strong relaxation (see Fig. 3.29): the phosphorus surface atoms move outwards whereas indium atoms relax inwards. In this manner, the phosphorus atoms approach the conventional p^3 bonding configuration while the indium atoms tend to sp^2 planar bonding (Fig. 3.17). The strength of surface-related vibrational features seen in the Raman spectrum is considerably weaker for a clean surface than for an Sb-covered one. Surface features can, however, be identified by measuring the spectra of a clean and a subsequently oxidized surface (Fig. 3.30). One attributes to modes of the clean surface the features that disappear upon oxidation. These features exhibit resonances at photon energies (2.5, 2.8 eV) well separated from those of the bulk (e.g., the E_1 gap at 3.1 eV), a fact that is responsible for their appearance in the Raman spectra with sufficient strength to be discriminated from the bulk signal.

Section 3.6 discusses *microscopic* surface phonons at interfaces between III–V compounds. These modes appear in cases in which the two bulk constituents have no common atom, e.g., an InSb/GaAs interface. Depending on the growth sequence, two types of interfaces are found, consisting of either a Ga–Sb or an In–As double layer. Correspondingly, GaSb– and InAs–like vibrations are observed.

The resonance effects utilized for the enhancement and observation of surface vibrations in Raman spectroscopy are similar to a number of other well-known Raman phenomena. Among them we mention first the so-called Surface Enhanced Raman Scattering (SERS) [1.21, 1.23]. In SERS, organic molecules adsorbed on metal surfaces (usually silver) exhibit enhancements in Raman cross sections by factors as high as 10^6. Of this factor, about 10^3

is believed to be due to modulation of the electronic structure of the adsorbing surface (similar to the resonance effects discussed above) while another 10^3 is due to electromagnetic resonances related to surface roughness. Electromagnetic resonances can also be used to enhance Raman scattering cross sections by placing the sample within a structure that acts as a Fabry-Perot resonator [1.24, 1.25]. Examples will be given, in Sect. 1.3. Here we mention only that by varying the angle of incidence of the laser on the resonantor it is possible to obtain a simultaneous double electromagnetic resonance, one involving the incident and the other the scattering photon.

1.1.3 Chapter 4

Chapter 4 of this volume, by T. Dekorsy, G.C. Cho, and H. Kurz, is concerned with the generation and detection of coherent phonons using subpicosecond (≈50 fs pulse width) lasers. These effects are closely related to conventional Raman spectroscopy in that

(1) excitations with $\boldsymbol{k} \simeq 0$ are involved,
(2) the detection and often also the excitation mechanism is related quantitatively to the Raman tensor of the phonons involved.

The technique is based on "zapping" a sample with a strong laser pulse with a width of ≃50 fs. The presence at the focal spot of a strong electromagnetic field produces large atomic displacements which go over into vibrations at the frequencies of phonon normal modes with $\boldsymbol{k} = 0$. The term *coherent* applies to these phonons because of their large amplitudes for a single mode, much larger than those of each of the phonons generated by conventional techniques (i.e., spontaneous Raman scattering, thermal excitation, etc.). Actually the latter, *incoherent phonons*, have individual amplitudes that tend to zero with increasing volume, as opposed to the large amplitudes of the coherent phonons under consideration.

After a large-amplitude coherent phonon has been generated, its vibrations with a period larger than the width of a laser pulse modulate the dielectric function and, correspondingly, the reflectivity and transmissivity of the crystal (the latter is of interest only in the case of transparent crystals). The mechanism of modulation is closely related to the Raman tensor, i.e., to the derivative of the dielectric function with respect to the phonon displacement (assuming, for simplicity, that the phonon frequency is smaller than the relevant widths of electronic states). For opaque materials, such as high-T_c superconductors and semiconductors in the visible range, the coherent phonons must be detected by reflection. The observed time dependence of the reflected pulse can be represented by a steeply rising and slowly decreasing background (see Fig. 4.8) onto which a small modulation, with the period of the phonon, is seen. The oscillating signal vs. time, obtained after subtracting the background (see lower panels in Fig. 4.8), contains the phonon spectrum in the time domain. Fourier transformation yields a frequency-dependent spectrum equivalent to that found directly in the frequency domain by standard spontaneous Raman spectroscopy. The time-domain measurements (using coherent phonons) are particularly suitable for long vibrational periods (i.e., low frequencies) while the conventional Raman spectra measured in the frequency domain are more appropriate for short periods, i.e., high frequencies. In some sense, both techniques are complementary. The decay of the coherent-phonon oscillations contains information about phonon lifetimes (i.e., linewidth in frequency space). Their phase contains information about the phonon generation mechanism.

In most coherent-phonon experiments the phonons are created by a laser pulse of the same train as that used for the detection. The latter is delayed with respect to the former by means of a standard delay mechanism. It is therefore difficult to ascertain the phonon creation mechanism since it is inexorably tied to the detection process. In the case of detection by reflection, the change in the reflectivity R due to the presence of the coherent phonon is

$$\frac{\Delta R(\omega)}{R} = \left(\beta_\mathrm{r} \frac{\mathrm{d}\epsilon_\mathrm{r}(\omega)}{\mathrm{d}u} + \beta_\mathrm{i} \frac{\mathrm{d}\epsilon_\mathrm{i}(\omega)}{\mathrm{d}u} \right) u \,, \tag{1.1}$$

where $\epsilon(\omega) = \epsilon_\mathrm{r} + \mathrm{i}\epsilon_\mathrm{i}$ is the frequency-dependent dielectric function, u the phonon amplitude and β_r, β_i the so-called Seraphin coefficients [1.26] that have been profusely used in the field of optical modulation spectroscopy [1.27]; $\mathrm{d}\epsilon_\mathrm{r}(\omega)/\mathrm{d}u$ and $\mathrm{d}\epsilon_\mathrm{i}(\omega)/\mathrm{d}u$ are the frequency-dependent components of the Raman tensor which include resonance effects. The relationship between resonant Raman scattering and modulation spectroscopy has been pointed out earlier [1.28].

Equation (1.1) can, in principle, be used to interpret the resonant behavior of the coherent phonon signal $\Delta R(\omega)/R$. In doing so, we must take into account the fact that u, the amplitude of a given coherent phonon, is also a function of the frequency ω of the generating laser. Two types of generating mechanisms have been proposed. One of them, designated as impulsive excitation, is also based directly on the Raman tensor which allows us to write down schematically the following contribution to the total free energy

$$\Delta F = \frac{E^2}{8\pi} \frac{\mathrm{d}\epsilon}{\mathrm{d}u} u \,, \tag{1.2}$$

where E is the electric field associated with the laser pulse. While the pulse is present, (1.2) leads to a finite atomic displacement u with the sign required to minimize ΔF. This mechanism of generation of coherent phonons is called "impulsive generation" [1.29]. The other mechanism is referred to as *displacive mechanism.* It obtains, for instance, when the laser pulse produces electron–hole pairs that remain after the pulse is gone, until they recombine. The presence of the photoinduced carriers produces a quasistatic phonon displacement u, that is accompanied by dampened vibrations at the frequency of the coherent phonons.

Sections 4.2 and 4.3 discuss the generation and detection of coherent phonons, respectively. Section 4.4 presents observations of coherent LO phonons in bulk undoped GaAs and discusses the various generation mechanisms applicable to this case. It also presents similar observations of coherent excitations involving plasmon–LO-phonon mixed modes.

Section 4.5 discusses coherent excitations in GaAs/AlSb superlattices, including the observation of Bloch oscillations. Section 4.6.1 discusses coherent phonons in tellurium, whereas in Sect. 4.6.2 terahertz emission related to the generation of coherent phonons is presented. Section 4.6.3 deals with the applications of the coherent phonon techniques to the investigation of the dynamics of these transitions. Finally, Sects. 4.7 and 4.8 discuss investigations of coherent phonons in a wide range of other materials, including fullerenes, ferroelectrics, high-T_c superconductors (see Fig. 4.24), and quantum dots. Of particular interest are the changes in phonon parameters, especially their scattering efficiency, when crossing T_c in high-T_c superconductors, the so-called phonon anomalies that are also observed in inelastic Raman scattering [1.30]. A quantitative connection between the two different observations of these anomalies has not yet been made.

1.2 Selected Recent Developments

The developments in the field of light scattering in solids since the appearance of Vol. VI of this series in 1991 have been so numerous that we must confine ourselves to a few highlights in which we have been particularly interested. Developments related to the contents of *Light Scattering in Solids VII*, having appeared recently, have already been reviewed in the introduction to that volume.

1.2.1 Resonant Raman Enhancement at Microcavities

Advances in microfabrication nowadays allow the placement of a Raman sample (a thin slab) in an electromagnetic microcavity by cladding both sides of the slab with so-called Distributed Bragg Reflectors (DBR, see Fig. 1.2). When either the laser or the scattered frequency equals that of a cavity resonance mode, a so-called single resonance is obtained. Exact resonance of one of these frequencies can be set by using a structure with a slight taper (a wedge): by moving the laser spot along the tapered direction, surface points at which single resonance conditions hold can be found [1.25]. This can also be accomplished by means of a tunable laser. Under the assumption that resonance for the scattered beam has been achieved, it is possible to obtain an *additional* resonance with a cavity mode for the incident beam (i.e., double resonance) by varying the angle of incidence Θ (outside of the sample) of the laser beam with respect to the normal to the sample surface. Under these conditions, double resonance takes place when the following condition holds:

$$\omega_L = \omega_s / \cos(\Theta / n_{\text{eff}}) \tag{1.3}$$

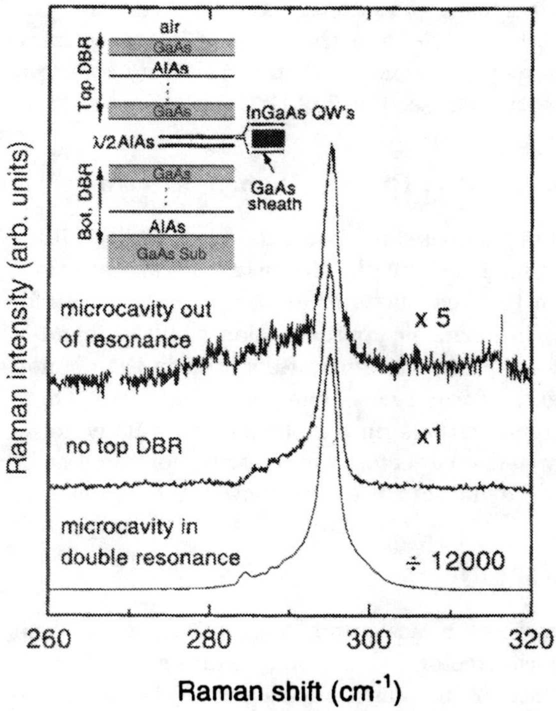

Fig. 1.2. Raman spectra taken in $z(x,y)\bar{z}$ configuration for a cavity such as that sketched in the inset and for a similar nanostructure without the top Distributed Bragg Reflector (DBR); Θ represents the angle of the incident beam with the superlattice axis; for incoming resonance with a cavity mode $\Theta = 54°$ using 1.37 eV laser photon energy. Note the enhancement, for double resonance with microcavity modes, by a factor of 12000 [1.24]

where n_{eff} is an average refractive index of the microstructure and Θ is assumed to be small. Figure 1.2 shows, beside a sketch of the microstructure, three spectra that correspond to GaAs-like LO-phonons of the $In_{0.14}Ga_{0.86}As$ quantum wells. The spectrum labeld "no top DBR", given for reference, corresponds to a reference sample grown without resonant cavity. The two other spectra were taken for the resonant structure; the lower one, 12000 times stronger than the peak of the "no top DBR" structure, was obtained under double-resonant conditions, whereas the peak in the upper curve (5× larger than for the non-resonant structure) was obtained off resonance [1.24]. Resonance spectra of the so-called interface modes (see below) have also been observed under resonant conditions (see Fig. 4 of [1.24]).

The spectra of Fig. 1.2 were taken under resonant conditions involving only the microcavity. It is also possible to explore resonant Raman scattering in situations where not only incident and scattered modes are resonant with cavity modes but also resonance with the intermediate states (excitons) takes place [1.31]. To obtain these resonant conditions, the laser frequency, the position of the laser spot on the sample, and the angle of incidence, may have to be varied. One refers to this "triple" resonance as *cavity polariton* mediated resonance Raman scattering (see Fig. 2 of [1.31]).

1.2.2 Effects of Anharmonicity on Phonon Raman Spectra

The first-order optical spectra of phonons should be delta-function-like within the harmonic approximation, broadened only by the instrumental resolution. Such delta functions follow, in the frequency domain, from the infinite lifetime of the phonons. Actually the harmonic approximation is only an *approximation*: anharmonicity is always present and manifests itself in the phonon widths, corresponding, e.g., to the decay time of one phonon into two. These widths are finite even at low temperatures since a phonon can usually decay into two or more with energy and wavevector conservation. For the case of one phonon of frequency ω_0 decaying into two, the linewidth $\Gamma(\omega_0)$ can be written as

$$\Gamma(\omega_0) = \Gamma_0 \left[1 + n_{\mathrm{B}}(\omega_1) + n_{\mathrm{B}}(\omega_2)\right] , \tag{1.4}$$

where n_{B} represents the Bose–Einstein factor and $\omega_1 + \omega_2 = \omega_0$; $\omega_1(\boldsymbol{k}_1)$ and $\omega_2(\boldsymbol{k}_2)$ are also related by the wavevector conservation condition $\boldsymbol{k}_1 + \boldsymbol{k}_2 \simeq 0$. Equation (1.4) can be simplified by assuming that $\omega_1 \simeq \omega_2$ ($= \omega_0/2$). In this case (1.4) contains only one adjustable parameter: the anharmonic low-temperature linewidth.

Considerable progress in the understanding of anharmonic phonon linewidhts has been made in the past three years thanks to the possibility of calculating *ab initio* the anharmonic force constants using *density functional perturbation theory* [1.32]. The conjecture expressed earlier [1.34] that for the Raman phonons of Ge and Si the observed temperature dependence of $\Gamma(\omega_0)$ requires that $\omega_1 \simeq 2\omega_2$, has been fully confirmed by the theoretical calculations [1.32].

Within the language of many-body theory, the linewidth Γ (FWHM) is related to the imaginary part of a *self energy* Σ ($\Gamma = -2\Sigma_{\mathrm{i}}$). The real part of Σ, Σ_{r}, introduces a temperature-dependent anharmonic shift of the phonon frequency; Σ_{r} thus represents the "third-order" anharmonicity correction to ω_0. Besides this third-order correction, there is a fourth-order contribution to the anharmonic shift of ω_0. Consequently, this shift is more difficult to calculate than the phonon width. First *ab initio* results can be found in [1.35]. We show in Fig. 1.3 the temperature dependence of the TO frequencies of GaAs, measured by Raman spectroscopy [1.33] compared with the results of *ab initio* calculations [1.35]. The agreement is excellent. The total shift is composed of the contributions of third-order anharmonicity, fourth-order anharmonicity, and thermal expansion, with the approximate relative magnitudes (3:1:1) and usually the same sign.

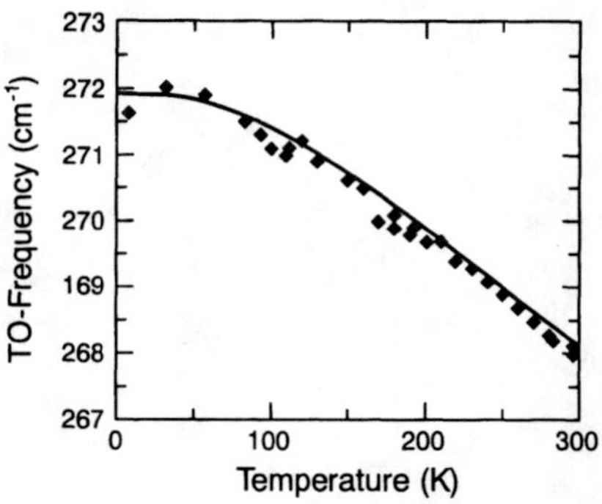

Fig. 1.3. The frequency shift of the TO phonon of GaAs vs temperature obtained by summing the various anharmonic contributions (thermal expansion, third-order, and fourth-order anharmonicity) compared with the measured values (data from [1.35])

The anharmonic widths (and shifts) just discussed are usually rather small, since anharmonic coupling constants are weak and the corresponding density of two-phonon states is usually also small. Under special circumstances, e.g., when ω_0 happens to lie very close to a critical point (van-Hove singularity) of the $\omega_1 + \omega_2$ two-phonon density of states, anharmonic effects can be very large and the typical Lorentzian lineshapes can become strongly distorted. This has been seen to be the case for the TO(Γ) phonons of CuCl [1.36] and also for its isoelectronic partner GaP [1.39]. These anharmonic effects can be "tuned" by varying either the isotopic masses (see Sect. 1.2.3) or by application of hydrostatic pressure which can lead to a shift of ω_0 with respect to the density of states corresponding to $\omega_1 + \omega_2$. We show in Fig. 1.4 the calculated and the measured dependence of the linewidth of the Γ phonons of silicon and germanium on hydrostatic pressure [1.37]. A small readjustment of the theory by hand was required to obtain nearly perfect agreement between calculated and measured linewidths vs pressure shown in Fig. 1.4.

1.2.3 Effect of Isotopic Composition on the Raman Spectra of Phonons in Semiconductors

After the fall of the "Iron Curtain" stable isotopes of many elements, made in Russia, became available in the West at affordable prices or even gratis, on the basis of scientific collaboration. The competition put the prices of other suppliers (mainly in the USA) under pressure. Since then, bulk single crystals and single-crystalline thin layers with different isotopic compositions have been grown at several laboratories, using these isotopes as starting materials. Raman spectroscopy has become an important technique for characterizing these materials, especially the nanostructures (e.g., superlattices) built from them since they cannot be investigated by x-ray diffraction.

The application of Raman spectroscopy to crystals with different isotopic compositions has generated a large amount of knowledge concerning the following properties of phonons [1.40]:

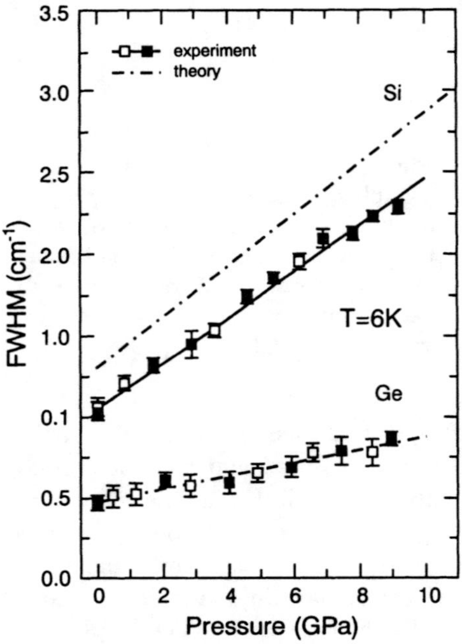

Fig. 1.4. Dependence of the linewidths of the Raman phonons of Si and Ge on hydrostatic pressure. The *symbols* represent experimental data, while the *dash-dotted lines* are the results of *ab initio* calculations. The *solid line* is a simple guide to the eye [1.37]

1) Phonon eigenvectors [1.38];
2) Anharmonic decay mechanisms; real and imaginary parts of the self-energy [1.41, 1.42];
3) Contribution of isotopic disorder to phonon linewidths: real and imaginary parts of the self-energies [1.41];
4) Forbidden scattering induced by isotopic disorder [1.43, 1.49].

Changes of isotopic composition lead to two different types of effects:

1) Effects related to changes in average mass (such as changes in lattice parameters [1.45], phonon frequencies, anharmonic properties, etc.). The anharmonic changes are related to changes in the vibrational amplitudes of the phonons. These amplitudes vary like $M^{-1/4}$ ($M \simeq$ isotopic mass) at low temperatures but are independent of M at high temperatures. Hence, these effects must be observed at low temperatures.
2) Effects related to fluctuations in the isotopic masses (for crystals with mixed isotopic composition). The largest of these effects is the decrease of thermal conductivity with increasing isotopic disorder [1.46]. It can amount to as much as two orders of magnitude.

Increases in linewidths induced by isotopic disorder, although much smaller than those found for the thermal conductivity, can also be observed by Ra-

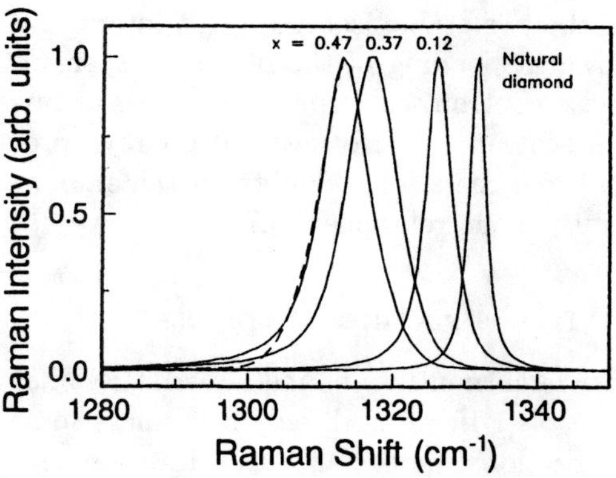

Fig. 1.5. Spectra of the Raman phonons of diamond with different isotopic composition ($^{12}C_{1-x}{}^{13}C_x$). Notice the broadening due to isotopic disorder. The peak found for $x = 0.47$ is asymmetric. The *dashed line* represents the low-frequency flank that would correspond to a symmetric line. From [1.43]

man spectroscopy for $\boldsymbol{k} \simeq 0$ Raman-active phonons. Note that in polyatomic crystals the effects of varying the isotopic mass of different atoms may be different [1.47].

The effect of isotopic disorder on the linewidth is very small for the highest frequency phonon. Such effect results from *elastic scattering* by the isotopic mass fluctuations. In the case of the highest phonon, no final states exist for scattering elastically into them, hence the effect should be zero within the first Born approximation. Higher order effects, involving anharmonicity and isotope disorder scattering, both to second order, produce a minor increase in such linewidth, as small as 0.02 cm^{-1} for the Raman phonons of germanium, for which the anharmonic linewidth at $T \approx 0$ is 0.7 cm^{-1} (FWHM) [1.48].

The nearly negligible isotopic broadening is a specific property of the highest frequency phonons, i.e., the Raman phonons of Si, Ge, gray tin, and the LO phonons of most III–V zincblende-type semiconductors. It does not apply to phonons at a generic point in the Brillouin zone. The optical branches of such generic phonons exhibit isotopic broadenings of the order of 1 cm^{-1} [1.49]. This broadening can be represented by a temperature-independent additive constant in (1.4). In polar semiconductors (e.g., GaAs, ZnSe, etc.) the Γ phonons split into an LO singlet and a TO doublet. The singlet is usually the highest and, as such, is barely broadened by isotopic fluctuations. The TO phonons at Γ are lower in frequency and are often degenerate with phonons from the same or other (LO) branches. Therefore the TO phonons experience considerable broadening when confronted with isotopic mass fluctuations [1.50].

Another particularly interesting case is that of diamond [1.43, 1.51]. As shown in Fig. 1.5, the Raman spectra of diamond broaden strongly (by 4 cm^{-1}, FWHM) when changing the natural isotopic composition ($^{12}C_{0.99}{}^{13}C_{0.01}$)

into nearly equal concentrations of the two stable isotopes. This interesting fact has been attributed to the upwards bending of the phonon dispersion relations around the Γ-point. Inelastic neutron scattering has confirmed this conjecture [1.52]. However, Raman scattering performed with x-rays produced by an electron-storage ring[3] [1.53] casts some doubts on the amount of upwards bending of the phonon dispersion relations of diamond.

1.2.4 Superlattices and Other Nanostructures: Phonons

Since the work on magnetic excitations in multilayers and superlattices has been thoroughly discussed in Chap. 3 of [1.1], we shall only cover here some recent highlights in scattering by vibrational modes and, to a lesser extent, by electronic excitations. Because of the direct connection with Sect. 1.2.2 we start our discussion with isotopic superlattices.

It is a "dogma" of light scattering in solids that if lasers around the visible region are used, only excitations with the $\boldsymbol{k}$ vector near Γ contribute to first-order scattering (this restriction is lifted, however, if strongly monochromatized x-rays are used as a source [1.53]). An isotopic superlattice [1.44] introduces a supercell and a mini-Brillouin-zone (mini-BZ). Although $\boldsymbol{k}$ must still be conserved, the $\boldsymbol{k} \simeq 0$ condition applies now to the mini-Brillouin-zone and therefore a much larger number of *folded* Raman-active phonons appear than for the corresponding bulk crystal. These modes can be mapped on the dispersion relations of the phonons in a mono-isotopic (or randomly isotopically mixed) bulk crystal [1.44, 1.54].

We now switch to light scattering by superlattices[4]. It is well known [1.20] that acoustic phonons in bulk crystals can be observed by means of *Brillouin* spectroscopy [1.55]. In superlattices composed of different semiconductors, the folding of the Brillouin zone introduces a number of additional, acoustic-like, states in the mini-zone that become accessible to Raman (or Brillouin) spectroscopy [1.20, 1.57]. During the past 10 years, a number of novel effects have been observed concerning these folded and other acoustic phonons under conditions of resonance with the lowest exciton of the wells [1.58]. Most of these phenomena follow from fluctuations in layer thickness, mainly (a) along the axis of growth (i.e., from layer to layer) but also (b) in-plane. The (a)-type fluctuations induce a random variation of the energy of electronic resonant transitions (gaps) of the otherwise equivalent layers (usually the so-called *wells*, GaAs in GaAs/AlAs systems) leading to a breakdown of the translational invariance along the superlattice axis (z-axis) and the concomitant breakdown of k_z conservation. The fluctuation in the gaps, however, is

[3] Synchrotron radiation; the possibility of performing phonon Raman scattering with x-rays is one of the most exciting developments in the field of the past five years.

[4] Because of differences in electronic properties one often distinguishes between superlattices, forming minibands, and multiple quantum wells, in which the lowest electronic states do not couple for $\boldsymbol{k}_\perp = 0$.

rather small, resulting typically from monolayer thickness fluctuations. For laser frequencies far away from resonance, the gap fluctuations are negligible and the nanostructure behaves as a perfect superlattice: sharp lines correspond to Brillouin scattering and also scattering by folded acoustic phonons appears. When the laser frequency approaches that of the electronic resonance ω_g the quantity $\omega_L - \omega_g$ that enters into the energy denominators of the standard expressions for the scattering efficiency approaches zero: small random fluctuations in ω_g result in large *relative* fluctuations in $\omega_L - \omega_g$. These fluctuations introduce a large incoherent component in the light scattering by acoustic phonons (without k_z conservation) that manifests itself in a broad light emission background below the laser line (see Fig. 1.6). Such background has been sometimes referrred to as luminescence". It results from a well-defined process involving the emission of the scattered photon plus an acoustic phonon (without k_z conservation) because of the interlayer thickness fluctuations but the in-plane $\boldsymbol{k}_\perp$ being conserved if the layers are flat. It is therefore more appropriate to designate the background as Raman scattering involving *one* acoustic phonon. Sometimes the term *geminate recombination* (gemini = twins) is also used.

Figure 1.6a shows a resonant Raman scattering spectrum that displays the broad background just mentioned. Since the sample still has an average translational period along z, a spectral component corresponding to k_z (and $\boldsymbol{k}_\perp$) conservation must also be present. It manifests itself as the weak Brillouin peak, labeled B in Fig. 1.6 of the average GaAs/AlAs bulk material, and the two folded LA phonons labeled FLA.

The structures just mentioned dominate in the off-resonance spectra. As shown in Fig. 1.6a, they become rather weak in the resonant spectrum that exhibits, as dominant features, complex structures that can be related to the frequencies of LA ($*$) and TA ($\bullet$) modes at the edge of the mini-BZ. The structures labeled ▼ in Fig. 1.6a,b have been assigned to anticrossings of the folded LA and TA branches [1.58]. Although these branches have different symmetries for $\boldsymbol{k}_\perp = 0$, and therefore should not anticross along k_z, they will anticross at finite $\boldsymbol{k}_\perp$, such as the $\boldsymbol{k}_\perp$ that must be included if in-plane thickness fluctuations are possible (see Fig. 1.6c). The Raman spectrum calculated for a single slab (so as to simulate k_z non-conservation) is shown in Fig. 1.6b. Figure 1.6d illustrates the effect of the non-conservation of $\boldsymbol{k}_\perp$ in bringing out into the spectra the frequencies of LA–TA anticrossings. Such anticrossings, forbidden by symmetry for $\boldsymbol{k}_\perp$, have nevertheless been alleged to occur for [100] superlattices in [1.59]. A close look at Fig. 4 of [1.59], however, indicates that the apparent disagreement with group theory is due to an error produced by the plotting software (authors, and also journal editors, should beware of such errors!). For similar superlattices grown along directions other than [100] or [111] the symmetry is lower and LO- and TO-like branches can anticross [1.60].

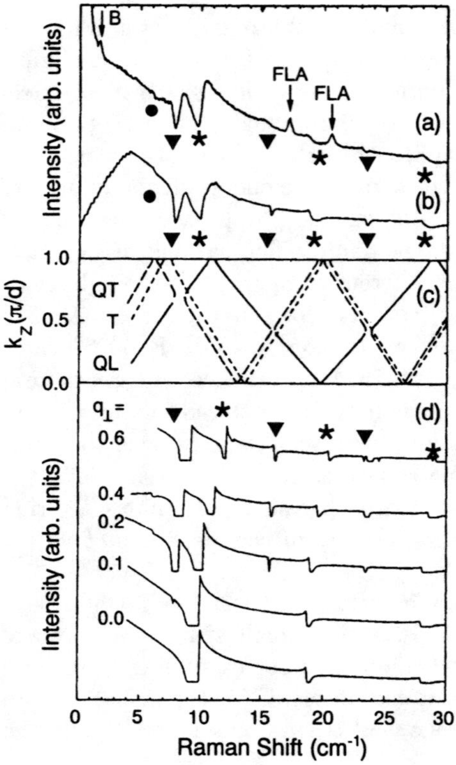

Fig. 1.6. (**a**) Experimental Raman spectrum of a (16/16) monolayer GaAs–AlAs superlattice compared to (**b**) the best theoretical profile obtained by integration over a range of in-plane crystal momenta $\boldsymbol{k}_\perp$. *Stars*, *triangles*, and the *circles* denote peaks and dips due to LA, LA-TA internal, and TA dispersion gaps, respectively. (**c**) Folded phonon dispersion vs k_z calculated for a nonzero in-plane wavevector $k_\perp = 0.4\pi/d$. *Solid*, *long-dashed*, and *short-dashed* lines indicate the dispersion branches of quasilongitudinal (QL), quasitransverse (QT), and pure transverse (T) modes. Zone-edge and internal gaps give rise to the intensity anomalies denoted in (**b**). (**d**) Theoretical spectra calculated for various values of k_z (given in units of π/d). For details, and for the parameters used in calculation, see [1.58]

We have just discussed recent advances concerning Raman spectroscopy of folded acoustic phonons in superlattices. As already treated in [1.20], the optical phonons of a superlattice also exhibit a number of rather interesting properties, among them the possibility of mapping the full dispersion relation of the constituent bulk materials. In this respect, the case of isotopic superlattice has already been mentioned in Sect. 1.2.3.

When a slab of a given material (e.g., GaAs) is sandwiched between two slabs of another material with rather different dispersion relations (i.e., little overlap between the optical phonon bands, e.g., GaAs/AlAs, see Fig. 31 of

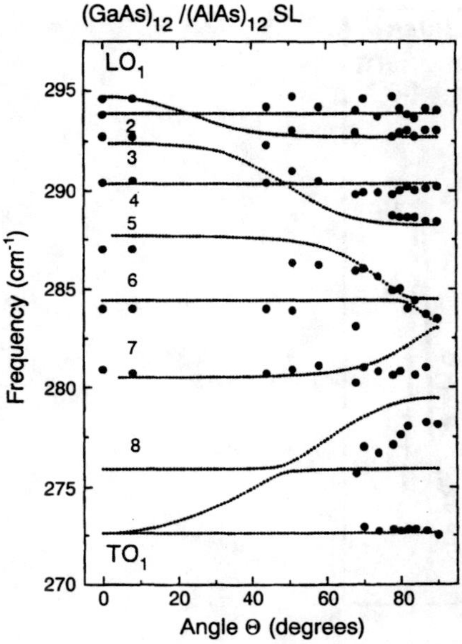

Fig. 1.7. Theoretical (*lines*) and experimental (*circles*) results for the angular dispersion of the GaAs-like optical modes of a $(GaAs)_{12}/(AlAs)_{12}$ superlattice. The calculation is for $|\boldsymbol{k}| = 7.8 \times 10^5\ \mathrm{cm}^{-1}$ and $\Theta = k_\perp / k_z$ [1.61]

[1.20]) *confined phonons* arise. The optical phonons of either component are "confined" to the individual slabs (this does not happen in general for acoustic phonons, for which overlap, at least partial, always takes place). The phonon amplitudes u obey one of the two approximate laws:

$$u \sim \cos \frac{m\pi}{d} z; \quad m = 1, 3, 5, \dots\ , \tag{1.5}$$

$$u \sim \sin \frac{m\pi}{d} z; \quad m = 2, 4, 6, \dots\ . \tag{1.6}$$

In (1.5) the origin of z has been taken to be at the center of the slab under consideration (of thickness d). For a [100] GaAs/AlAs superlattice the point group is D_{2d} and the z-polarized modes have, within this group, symmetries B_2 (for $m = 1, 3, 5, \dots$) and A_1 (for $m = 2, 4, 6, \dots$). Phonons of either B_2 or A_1 symmetry are both Raman allowed in the superlattice, although A_1 is forbidden in bulk materials. The B_2 modes are also ir-active, whereas the A_1 modes are not.

Beside the confined, bulk-like modes of (1.5) one finds in these superlattices electrostatic macroscopic interface modes of the type discussed in Sect. 1.2.3. These modes have a finite in-plane propagation vector $k_\perp \gtrsim k_z$. Since they result from the polar nature of the bulk constituents (i.e., their ir-activity) they should mix with the ir-active modes of (1.5). Macroscopic methods can be used to calculate the dispersion relations that combine interface electrostatic properties with the bulk confinement [1.19, 1.61]. The $m = 1$ mode of (1.5) turns out to be basically the so-called electrostatic interface mode (the introduction of ir-activity cannot increase the number of modes) for small values of $k_\perp / k_z$ and anticrosses with the $m = 3$ and

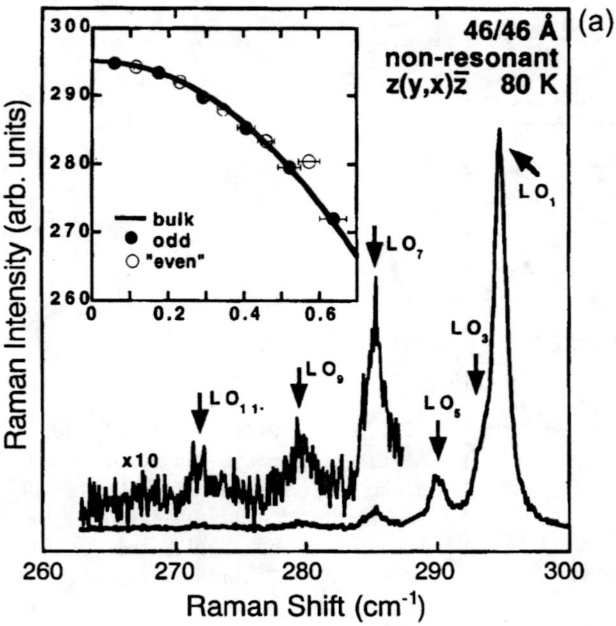

Fig. 1.8. (**a**) Raman spectrum taken on a 46/46 Å GaAs/AlAs MQW at 80 K in $\bar{z}(x,y)z$ polarization, with 2.54 eV laser line, corresponding to nonresonant conditions. The peaks are due to odd-order confined modes of B_2 symmetry. The frequencies of these peaks are plotted against their effective confinement $\boldsymbol{k}$ vector (*solid symbols*) in the inset; they map the bulk GaAs LO dispersion (*solid lines*). Also plotted (*open symbols*) are the frequencies of the peaks in the outgoing resonant spectrum, assuming (incorrectly, see Fig. 1.8b, opposite page) that they are due to even-order modes [1.62]

$m = 5$ modes with increasing $k_\perp/k_z$, thus confirming the ir-active nature of the modes (1.5). The (1.6) modes, on the contrary, do not interact with the ir-active modes mentioned above. Figure 1.7 displays the frequencies obtained from Raman measurements on a $(GaAs)_{12}/(AlAs)_{12}$ superlattice vs the angle Θ between the z-axis and the backscattering direction of the light ($\tan\Theta = k_\perp/k_z$). The dotted curves were calculated with the macroscopic method discussed in [1.19].

Concerning the polarization selection rules, the same as those of the bulk materials apply to the B_2 phonons of (1.5) in the off-resonance case $[\bar{z}(x,y)z]$. This is clearly seen in Fig. 1.8a [1.62]. The A_{1g} phonons of (1.6) can also be seen off-resonance in the $[\bar{z}(x,x)z]$ or $[\bar{z}(y,y)z]$ configurations. Near resonance, however, the A_1-like spectra change shape and develop into a series of oscillations of nearly equal amplitude, except for the strong $m \simeq 2$ peak. In Fig. 1.8b, five such oscillations are seen. Their minima have been shown to correspond to the gaps that appear in Fig. 1.7 between the $m = 3, 5, \ldots$

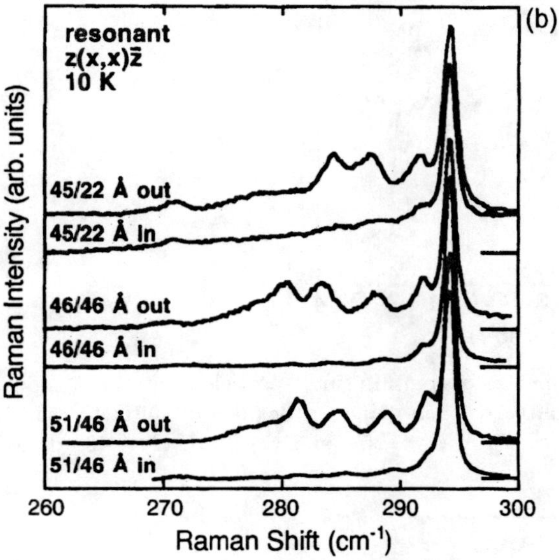

Fig. 1.8. (**b**) Comparison of the GaAs optic phonon region of Raman spectra taken for the outgoing and incoming resonance conditions. Spectra taken on MQWs with different layer dimensions (as indicated) are plotted, each recorded for $\bar{z}(x,x)z$ polarization at 10 K. The spectra are normalized to have the same height and shifted vertically for clarity [1.62]

B_2 modes when anticrossing the $m = 1$ interface mode [1.62]. For a detailed recent discussion of Raman scattering by phonons in superlattices see [1.63].

Before closing this review of highlights in Raman scattering by superlattices and quantum wells we mention one interesting development involving excitations of conduction electrons in GaAs/AlAs that has taken place after [1.64] was written. As discussed in [1.64], two types of excitations are possible, charge-density excitations,indexexcitations!spin-density obtained for parallel incident and scattered polarizations, and spin-density excitations (spin flip) involving perpendicular polarizations. In a GaAs/AlAs quantum well the lowest conduction band is spin-split according to:

$$\Delta E_{\mathrm{c}}(\boldsymbol{k}_\perp) = \pm\gamma \left[\kappa^4 k_\perp^2 - \left(4\kappa^2 - k_\perp^2\right) k_x k_y\right]^{1/2} , \tag{1.7}$$

where $\kappa \simeq \pi/d$, $\boldsymbol{k}_\perp$ (components k_x and k_y) the in-plane wavevector of the electron, and γ is a bulk band-structure parameter (for GaAs $\gamma \simeq 20$ eVÅ^3) [1.65, 1.66].

Figure 1.9 shows the spin-flip spectra of a GaAs/$Al_{0.33}Ga_{0.67}As$ quantum well doped with 1.3×10^{12} electrons/cm^2, for an in-plane scattering vector $\Delta k_\perp = 0.49 \times 10^5$ cm^{-1} along the [010] and [011] in-plane direction and also along a direction at an angle of 25° with [010]. The splitting of the doublets in Fig. 1.9 corresponds to the spin-splitting of (1.7). Note that it is smaller in the [011] than in the [010] peak, thus reflecting the anisotropy imposed by the $k_x k_y$ terms in (1.7).

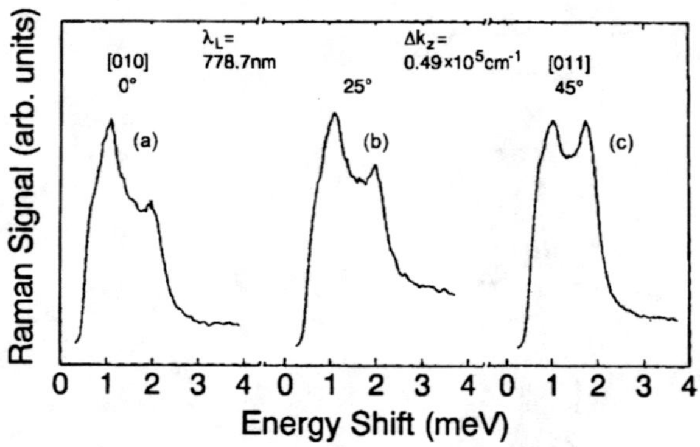

Fig. 1.9. Depolarized Raman spectra of spin-flip single-particle excitations, showing the inversion asymmetry splitting, for in-plane vector $\Delta k_{\perp} = 0.49 \times 10^5$ cm^{-1} along the (**a**) [010], $\Theta = 0°$, (**b**) [010], $\Theta = 25°$; and (**c**) [011], $\Theta = 45°$ directions [1.66]

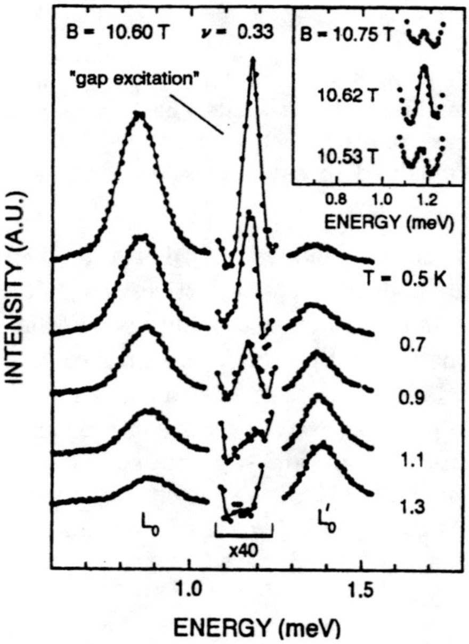

Fig. 1.10. Temperature dependence of inelastic light scattering spectra of a low-lying excitation of the FQHE at $\nu = 1/3$. The single quantum well has density $n = 8.5 \times 10^{10}$ cm^{-2}. The inset shows the B dependence of the 0.5 K spectra. The light-scattering peak labeled "gap excitation", is interpreted as a $q = 0$ collective gap excitation. The bands labeled L_0 and L_0' comprise the characteristic doublets of intrinsic photoluminescence [1.68]

Last but not least, we mention one of the most exciting recent developments in the field of electronic Raman scattering by electronic excitations in semiconductor nanostructures: scattering in the Fractional Quantum Hall Effect (FQHE) regime [1.67, 1.68, 1.69, 1.70, 1.71][5]. This effect is observed at very low temperatures ($\lesssim$1 K) in semiconductor quantum wells in a magnetic field, for electron concentrations such that the lowest Landau level is only partly occupied. By varying the magnetic field it is possible to fill only a fraction of the electronic state available in the first Landau level, the simplest cases being those of $\nu = 1/3$ and $\nu = 2/3$ occupation. Under these conditions, many-body effects induced in the two-dimensional electron gas (made zero-dimensional by the presence of the magnetic field) an excitation gap of the order of 1 meV (i.e., 12 K). This gap makes the many-particle electronic state incompressible at $T \simeq 0$. Light scattering, once again turns out to be an ideal technique for the observation of the gap. We show in Fig. 1.10 an example of such an observation: the sharp central peak, at $\simeq$ 1.2 meV, corresponds to scattering by excitations of the incompressible fluid; it disappears rapidly with increasing temperature. The broader peaks labeled L_0 and L_0' can be identified as luminescence associated with the recombination of electrons in the lowest Landau level with holes photocreated in the uppermost valence state [1.72]. The insert demonstrates that the 1.2 meV peak only appears at a very well defined magnetic field, that which corresponds to 1/3 filling the first Landau level.

In [1.68] it is shown that the gap excitations mentioned above do not only appear in the intra-Landau-level scattering but also, at higher energies, in scattering by inter-Landau-level excitations.

References

1.1 M. Cardona, G. Güntherodt (eds.): *Light Scattering in Solids VII: Crystal-Field and Magnetic Excitations* Topics Appl. Phys. **75** (Springer, Berlin, Heidelberg 1999)

1.2 M. Cardona, G. Güntherodt (eds.): *Light Scattering in Solids VI: Recent Results, Including High-T_c Superconductors*, Topics Appl. Phys. **68** (Springer, Berlin, Heidelberg 1991)

1.3 G. Schaack: Raman Scattering by Crystal-Field Transitions, in [1.1]

1.4 B. Hillebrands: Brillouin Scattering from Layered Magnetic Structures, in [1.1]

1.5 This quotation can also be found in *The Scientific Papers of C.V. Raman* (Indian Acad. of Sci., Bangalore 1978) p. 122

1.6 A. Smekal: Naturwiss. **11**, 873 (1923)

1.7 C.V. Raman: Ind. J. Phys. **2**, 387 (1928)
C.V. Raman, K.S. Krishnan: Nature **121**, 501 (1928)

[5] For the discovery in 1985 of the FQHE, R.B. Laughlin, H.L. Stormer, and D.C. Tsui were awarded the Nobel prize for physics in 1998

1.8 G. Landsberg, L. Mandelstam: Naturwiss. **16**, 557 (1928)
The date of submittal of this paper is May 6, 1928. In his inaugural address delivered to the South Indian Science Association in Bangalore on March 16, 1928, in which observations of inelastic light scattering for solids, liquids and gases are reported for the first time, Raman gives as the date of the first observation February 28, 1928. This address is reprinted in [1.5]
1.9 A.M. Heyns (ed.): Proc. XVIth Int (Conf. on Raman Spectrosocpy ICORS, J. Wiley, New York 1998)
1.10 B. Schrader (ed.): *Infrared and Raman Spectroscopy*, (VCH, Weinheim, New York 1995)
1.11 J. McHale: *R. Buckminster Fuller* (Braziller, New York 1962)
1.12 H.W. Kroto: Chem. Soc. Rev. **11**, 435 (1982)
1.13 A. Leger, L. d'Hendecourt, L. Verstraete, W. Schmidt: Astron Astrophys **203**, 145 (1987)
1.14 W. Krätschmer, D.R. Hufman: Chem. Phys. Lett. **170**, 167 (1990)
1.15 G. Els, P. Lemmens, P.H.M. van Loosdrecht, G. Güntherodt, H.P.Lang, V. Thommen-Geiser, H.-J. Güntherodt: Physica C **307**, 79 (1998)
1.16 H.D. Sun, Z.K. Tang, J. Chen, G. Li: Solid State Commun. **109**, 363 (1999)
1.17 M. Cardona, G. Güntherodt (eds.): *Light Scattering in Solids V: Superlattices and other Microstructures*, Topics Appl. Phys. **66** (Springer, Berlin, Heidelberg 1989)
1.18 M. Cardona, G. Güntherodt (eds.): *Light Scattering in Solids III: Recent Results*, Topics Appl. Phys. **51** (Springer, Berlin, Heidelberg 1982)
1.19 C. Trallero-Giner, R. Pérez-Alvarez, F. García-Moliner: *Long Wave Polar Modes in Semiconductor Heterostructures* (Pergamon, Oxford 1998)
1.20 B. Jusserand, M. Cardona: Raman Spectroscopy of Vibrations in Superlattices, in [1.17]
1.21 A. Otto: Surface-Enhanced Raman Scattering: Classical and Chemical Origins, in *Light Scattering in Solids IV: Electronic Scattering, Spin Effects, SERS, and Morphic Effects*, Topics Appl. Phys. **54** (Springer, Berlin, Heidelberg 1984)
1.22 M. Cardona, G. Güntherodt (eds.): *Light Scattering in Solids IV: Electronic Scattering, Spin Effects, SERS, and Morphic Effects*, Topics Appl. Phys. **54** (Springer, Berlin, Heidelberg 1984)
1.23 K. Arya, R. Zeyher: Theory of Surface Enhanced Raman Scattering. ibid, pp. 419
1.24 A. Fainstein, B. Jusserand, V. Thierry-Mieg: Phys. Rev. Lett. **75**, 3764 (1995)
1.25 A. Mlaya, O. Marco, J.R. Huntzinger, A. Zwick, R. Carles, V. Bardinal, C. Fontaine, A. Muñoz-Yagüe: J. Phys. Condens. Matter **10**, 9535 (1998)
1.26 B.O. Seraphin, N. Bottka: Phys. Rev. **145**, 628 (1966)
1.27 M. Cardona: In Solid State Physics, Suppl., ed. by Seitz, Turnbull, Ehrenreich (Academic, New York, 1969)
1.28 M. Cardona: Surf Sci **37**, 100 (1973)
1.29 R. Merlin: Solid State Commun. **102**, 207 (1997)
1.30 B. Friedl, C. Thomsen, H.-U. Habermeier, M. Cardona: Solid State Commun. **78**, 291 (1991)
1.31 A. Fainstein, B. Jusserand, V. Thierry-Mieg: Phys. Rev. Lett. **78**, 1576 (1997)
1.32 A. Debernardi, S. Baroni, E. Molinari: Phys. Rev. Lett. **75**, 1819 (1998)
1.33 G. Irmer, M. Wenzel, J. Monecke: Phys. stat. solidi (b) **195**, 85 (1996)

1.34 J. Menéndez, M. Cardona: Phys. Rev. B **29**, 2051 (1984)
1.35 A. Debernardi, M. Cardona: Physica B **263–264**, 687 (1999); A. Debernardi, Solid State Comm. Vol. 113, in press
1.36 A. Göbel, T. Ruf, Cheng-Tian Lin, M. Cardona, J.-C. Merle, M. Joucla: Phys. Rev. B **56**, 210 (1997)
1.37 C. Ulrich, E. Anastassakis, K. Syassen, A. Debernardi, M. Cardona: Phys. Rev. Lett. **78**, 1283 (1997)
1.38 F. Widulle, T. Ruf, O. Buresch, A. Debernardi, M. Cardona: Phys. Rev. Lett. **82**, 5281 (1999)
1.39 F. Widulle, T. Ruf, A. Göbel, E. Schönherr, M. Cardona: Phys. Rev. Lett., **82**, 3089 (1999)
1.40 M. Cardona: Physica B (Amsterdam) **263–264**, 376 (1999)
1.41 F. Widulle, T. Ruf, A. Göbel, I. Silier, E. Schönherr, M. Cardona, J. Camacho, A. Cantarero, W. Kriegseis, V.I. Ozhogin: Physica B **263-264**, 448 (1999)
1.42 D.T. Wang, A. Göbel, J. Zegenhagen, M. Cardona: Phys. Rev. B **56**, 13167 (1997)
1.43 J. Spitzer, P. Etchegoin, M. Cardona, T.R. Anthony, W.F. Banholzer: Solid State Commun. **88**, 509 (1993)
1.44 A. Göbel, T. Ruf, A. Fischer, K. Eberl, M. Cardona, J.P. Silveira, F. Briones: Phys. Rev. B, May 15 (1999)
1.45 A. Kazimirov, J. Zegenhagen, M. Cardona: Science **282**, 930 (1998)
1.46 M. Asen-Palmer, K. Bartkowski, E. Gmelin, M. Cardona, A.P. Zhernov, V.I. Ozhogin, A.V. Inyushkin, A. Taldenkov, K. Itoh, E.E. Haller: Phys. Rev. B **56**, 9431 (1997)
1.47 A. Göbel, T. Ruf, M. Cardona, C.T. Lin, J. Wrzeskinski, M. Steube, K. Reimann, J.-C. Merle, M. Joucla: Phys. Rev. B **57**, 15183 (1998)
1.48 J.M. Zhang, M. Giehler, A. Göbel, T. Ruf, M. Cardona, E.E. Haller, K. Itoh: Phys. Rev. B **57**, 1348 (1998)
1.49 A. Göbel, D.T. Wang, M. Cardona, L. Pintschovius, W. Reichardt, J. Kulda, N.M. Pyka, K. Itoh, E.E. Haller: Phys. Rev. B **58**, 10510 (1998)
1.50 A. Göbel, T. Ruf, J.M. Zhang, R. Lauck, M. Cardona: Phys. Rev. B **59**, 2749 (1999)
1.51 K.C. Haas, H.A. Tamor, T.R. Anthony, W.F. Banholzer: Phys. Rev. B **45**, 7171 (1992)
1.52 J. Kulda, B. Dorner, B. Roessli, H. Sterner, R. Bauer, Th. May, K. Karch, P. Pavone, D. Strauch: Solid State Commun. **99**, 799 (1996)
1.53 M. Schwoerer-Böhning, A.T. Macrander: Phys. Rev. Lett. **80**, 5572 (1998)
1.54 J. Spitzer, T. Ruf, M. Cardona, W. Dondl, R. Schorer, G. Abstreiter, E.E. Haller: Phys. Rev. Lett. **72**, 156l (1994)
1.55 A.S. Pine: Brillouin Scattering in Semiconductors, in [1.56]
1.56 M. Cardona (ed.): *Light Scattering in Solids I: Introductory Concepts*, 2nd edn., Topics Appl. Phys. **8** (Springer, Berlin, Heidelberg 1982)
1.57 T. Ruf, J. Spitzer, V.F. Sapega, V.I. Belitsky, M. Cardona, K. Ploog: *Raman Scattering by Acoustic Phonons in Semiconductor Superlattices* in Advances in Solid State Physics, ed. by R. Helbig (Vieweg, Braunschweig, 1994) p. 237
1.58 V.I. Belitsky, T. Ruf, J. Spitzer, M. Cardona: Phys. Rev. B **49**, 8263 (1994)
1.59 K. Lambert, G.P. Srivastava: Phys. Rev. B **56**, 1388 (1997)
1.60 F. Calle, M. Cardona, E. Richter, D. Strauch: Solid State Commun. **72**, 1153 (1989)

1.61 M. Zunke, R. Schorer, G. Abstreiter, W. Klein, G. Weimann, M.P. Chamberlain: Solid State Commun. **93**, 847 (1995)
1.62 A.J. Shields, M.P. Chamberlain, M. Cardona, K. Eberl: Phys. Rev. B **51**, 17728 (1995)
1.63 T. Ruf: *Phonon Raman Scattering in Semiconductors, Quantum Wells, and Superlattices*, Springer Tracts in Mod. Phys. **142** (Springer, Berlin, Heidelberg 1998)
1.64 A. Pinczuk, G. Abstreiter: *Spectroscopy of Free Carrier Excitations in Semiconductor Quantum Wells*, in [1.17]
1.65 P.V. Santos, M. Willatzen, M. Cardona, A. Cantarero: Phys. Rev. B **51**, 5121 (1995)
1.66 D. Richard, B. Jusserand, H. Peric, B. Etienne: Phys. Rev. B **47**, 16028 (1993)
1.67 R.E. Prange, S.M. Girvin (eds.): *The Quantum Hall Effect*, (Springer, New York 1990)
1.68 A. Pinczuk, B.S. Dennis, L.N. Pfeiffer, K. West: Phys. Rev. Lett. **70**, 3983 (1993)
1.69 H. D. M. Davies, J. C. Harris, J. F. Ryan, A. J. Turberfield: Phys. Rev. Lett. **78**, 4095 (1997)
1.70 C. Schüller, R. Krahne, G. Biese, C. Steinebach, E. Ulrichs, D. Heitmann: Phys. Rev. B **56**, 1037 (1997)
1.71 V.E. Kirpichev, I.V. Kukushkin, K. von Klitzing, K. Eberl: JETP Lett. **67**, 210 (1998)
1.72 D. Heiman, B.B. Goldber, A. Pinczuk, C.W. Tu, A.C. Gossard, J.H. English: Phys. Rev. Lett. **61**, 605 (1988)

2 Vibrational Spectroscopy of C_{60}

José Menéndez and John B. Page

With 14 Figures

The fullerene era was started in 1985 with the discovery of the stable C_{60} cluster and its interpretation as a cage structure with the familiar shape of a soccer ball [2.1]. An explosive growth in fullerene research was triggered in 1990 by the development of a method to produce fullerenes in bulk quantities [2.2]. Subsequently, the structural, electronic, and vibrational properties of many fullerenes have been studied in detail. The importance and potential of this new class of materials is exemplified by the discovery of intermediate-temperature superconductivity in doped C_{60} [2.3]. Carbon nanotubes, a novel form of carbon which combines the properties of graphite and fullerenes, were discovered in 1991 [2.4]. Because of their intriguing properties and potential for applications, nanotubes are currently the subject of very intense research. Polymerization of C_{60} molecules, particularly by photoexcitation [2.5] but by several other techniques as well, also results in a variety of interesting new structures, which contain both 4-coordinated and 3-coordinated carbon atoms [2.6]. The burgeoning field of fullerene research has been reviewed by several authors [2.7, 2.8, 2.9, 2.10, 2.11, 2.12], most notably by *Dresselhaus* et al. [2.11] in an extensive monograph that appeared in 1996.

In spite of the rapidly increasing interest in new forms of fullerenes, icosahedral C_{60}, the "most beautiful molecule" [2.13], remains the focus of vigorous research as the prototype fullerene system. The present chapter concerns the vibrational structure of C_{60} and the efforts to unravel its details using spectroscopic techniques. This remains a work in progress, but we hope to show that a look at the existing body of experimental and theoretical research from the broader perspective of an extended article provides a deeper understanding of the vibrational properties of C_{60}.

2.1 Vibrations in C_{60}

Several reviews have appeared on the vibrational structure of C_{60}, with emphasis on Raman and infrared spectroscopy [2.11,2.14,2.15]. We have tried to limit the overlap with these works by emphasizing recent developments and providing sufficient theoretical detail to facilitate the critical evaluation of comparisons between measured and calculated spectra. No attempt is made to

Topics in Applied Physics, Vol. 76
Light Scattering in Solids VIII
Eds.: M. Cardona, G. Güntherodt

provide complete bibliographic references on the topics discussed, as these are readily assembled, e.g., via the reviews cited above and by following up references in the work we discuss. Accordingly, we have tried to select those references which most usefully illustrate the points under consideration.

We begin in Sect. 2.1.1 with a brief introduction to relevant aspects of molecular vibrations, followed by a discussion in Sect. 2.1.2 of symmetry properties and spectroscopic selection rules for isolated icosahedral C_{60} molecules. Section 2.1.3 concerns symmetry-lowering perturbations arising from the intermolecular interactions in condensed phases and from the presence of carbon isotopes. Theoretical approaches used to calculate the vibrational frequencies and eigenvectors of C_{60} are discussed in Sect. 2.1.4. Also given there is our preferred frequency assignment for all of the vibrational modes in icosahedral C_{60}. The frequencies are based on data from a wide variety of vibrational spectroscopies and are compared with representative theoretical results. The experimental data are detailed in Sect. 2.2, with special emphasis on the spectroscopy of "silent" modes, i.e., those which are neither infrared nor Raman active in first order. Finally, Sects. 2.3 and 2.4 discuss infrared and Raman intensities. The development of accurate models to reproduce these intensities is an important prerequisite for the understanding of the much more complicated spectra of higher fullerenes, nanotubes and polymers.

2.1.1 Theoretical Basis

At a fundamental level, the vibrational properties of solids or molecules such as C_{60} derive from the many-electron states, as do the quantities which determine the spectroscopic coupling strengths, e.g., Raman polarizability derivatives and ir effective charges. The last decade has seen rapid advances in the development of numerous "first-principles" techniques for the practical calculation of electronic, vibrational, and structural properties from a unified many-electron point of view; some of these techniques will figure into our subsequent discussion.

For nonmetallic systems whose vibrational and electronic transition energies are well separated, the eigenstates may be described within the adiabatic (i.e., Born-Oppenheimer) approximation as products of coupled electronic and vibrational states

$$\Psi_{n\nu}(\boldsymbol{x},\boldsymbol{u}) = \phi_n(\boldsymbol{x},\boldsymbol{u})\psi_{n\nu}(\boldsymbol{u}), \tag{2.1}$$

where n and ν are the electronic and vibrational quantum numbers, respectively. The many-electron states $\phi_n(\boldsymbol{x},\boldsymbol{u})$ depend parametrically on the atoms' displacements, represented collectively by $\boldsymbol{u}$, and on the electronic coordinates $\boldsymbol{x}$. They are eigenfunctions of the many-electron problem for fixed atomic configuration $\boldsymbol{u}$:

$$[T_{\mathrm{E}} + V_{\mathrm{EE}}(\boldsymbol{x}) + V_{\mathrm{EN}}(\boldsymbol{x},\boldsymbol{u}) + V_{\mathrm{NN}}(\boldsymbol{u})]\phi_n(\boldsymbol{x},\boldsymbol{u}) = E_n(\boldsymbol{u})\phi_n(\boldsymbol{x},\boldsymbol{u}), \tag{2.2}$$

where T_{E} is the electronic kinetic energy operator and the nuclear-nuclear potential energy $V_{\mathrm{NN}}(\boldsymbol{u})$ has been included. The vibrations are then determined by taking the effective potential energy function for the atoms to be the many-electron ground state energy eigenvalue: $V(\boldsymbol{u}) \equiv E_0(\boldsymbol{u})$. This assumes that the system remains in its many-electron ground state as the atoms move.

The minimum of V defines the atomic equilibrium configuration. The atoms' displacements from equilibrium are given by the $3N \times 1$ configuration space vector $\boldsymbol{u} \equiv \{u(\ell\alpha)\}$, where $\ell = 1, \ldots, N$ labels sites and α denotes x, y, or z. Within the harmonic approximation, V is quadratic in the displacements:

$$V = \frac{1}{2} \sum_{\ell\alpha, \ell'\alpha'} \Phi(\ell\alpha, \ell'\alpha') u(\ell\alpha) u(\ell'\alpha') = \frac{1}{2} \tilde{\boldsymbol{u}} \boldsymbol{\Phi} \boldsymbol{u}, \tag{2.3}$$

where $\boldsymbol{\Phi} \equiv \{\Phi(\ell\alpha, \ell'\alpha') = [\partial^2 V / \partial u(\ell\alpha) \partial u(\ell'\alpha')]_0\}$ is the $3N \times 3N$ harmonic force constant matrix, with the derivatives evaluated at the equilibrium configuration $\boldsymbol{u} = \boldsymbol{0}$. The tilde denotes the transpose, and the zero of potential energy is taken at the equilibrium configuration. The vibrational Hamiltonian is then

$$H = \frac{1}{2}(\tilde{\boldsymbol{p}} \boldsymbol{M}^{-1} \boldsymbol{p} + \tilde{\boldsymbol{u}} \boldsymbol{\Phi} \boldsymbol{u}), \tag{2.4}$$

where the diagonal matrix $\boldsymbol{M} \equiv \{m(\ell) \delta_{\ell\ell'} \delta_{\alpha\alpha'}\}$ contains the atoms' masses and $\boldsymbol{p} \equiv \{p(\ell\alpha)\}$ contains their momenta.

The normal mode frequencies and displacement patterns are obtained by solving the eigenvalue problem

$$(\boldsymbol{\Phi} - \omega_f^2 \boldsymbol{M}) \boldsymbol{\chi}(f) = 0, \tag{2.5}$$

where $f = 1, \ldots, 3N$ labels the modes. Since both the force constant and mass matrices are real and symmetric, the eigenvectors may be taken as complete and orthonormal with respect to $\boldsymbol{M}$:

$$\sum_{f=1}^{3N} \boldsymbol{M} \boldsymbol{\chi}(f) \tilde{\boldsymbol{\chi}}(f) = \boldsymbol{I}, \tag{2.6}$$

and

$$\tilde{\boldsymbol{\chi}}(f) \boldsymbol{M} \boldsymbol{\chi}(f') = \delta_{ff'}. \tag{2.7}$$

The $3N$ normal coordinates $\{d_f\}$ and their conjugate momenta $\{p_f\}$ are linearly related to the atomic displacements and momenta through

$$\boldsymbol{u} = \sum_f \boldsymbol{\chi}(f) d_f \tag{2.8}$$

and

$$\boldsymbol{p} = \boldsymbol{M} \sum_f \boldsymbol{\chi}(f) p_f. \tag{2.9}$$

When these equations are substituted into (2.4) and the eigenvalue equation (2.5) and orthonormality condition (2.7) are used, the Hamiltonian reduces to that for a sum of independent harmonic oscillators

$$H = \frac{1}{2}\sum_{f}(p_f^2 + \omega_f^2 d_f^2). \tag{2.10}$$

The normal coordinate transformation is valid within either a classical or quantum mechanical treatment of the vibrational problem. In the former case the normal coordinates and their conjugate momenta satisfy Hamilton's equations, leading to $\ddot{d}_f + \omega_f^2 d_f = 0$. In the quantum case, these quantities are operators satisfying commutation relations $[d_f, p_{f'}] = \mathrm{i}\hbar\delta_{ff'}$, and the vibrational Hamiltonian may be reexpressed in the familiar form

$$H = \sum_{f} \hbar\omega_f(a_f^\dagger a_f + 1/2), \tag{2.11}$$

where $a_f \equiv (2\hbar\omega_f)^{-1/2}(p_f - \mathrm{i}\omega_f d_f)$ is the annihilation operator for mode f. The eigenenergies are then $E_{\{n_f\}} = \sum_f \hbar\omega_f(n_f + \frac{1}{2})$, with $n_f = 0, 1, 2, \ldots$ counting the number of quanta $\hbar\omega_f$ in mode f.

In either the quantum or classical description, the key physical quantities are the mode frequencies ω_f and corresponding displacement patterns $\boldsymbol{\chi}(f) = \{\chi(\ell\alpha|f)\}$, obtained by solving the eigenvalue problem (2.5). For fullerenes, one typically solves this problem directly, using force constants obtained from phenomenological potential models or from first-principles techniques; both types of calculations will play a role in this chapter.

2.1.2 Symmetry and Selection Rules

The high symmetry of an isolated, isotopically pure C_{60} molecule in its equilibrium configuration imposes very strong constraints on the form of the normal mode displacement patterns. All 60 sites are symmetry-equivalent, and the point group is the full icosahedral group I_h, consisting of 120 operations. These are the 60 proper symmetry operations of a regular icosahedron, together with each of these operations followed by the inversion. The mode displacement patterns are basis functions for irreducible representations of I_h; that is, they transform under a symmetry operation r as

$$\boldsymbol{\mathcal{R}}\boldsymbol{\chi}(n,i) = \sum_{j=1}^{l_n} \Gamma_{ji}^n(r)\boldsymbol{\chi}(n,j), \tag{2.12}$$

where the mode label has been split $(f) \to (n,i)$, with n labeling the irreducible representation and i denoting the partner within the representation. The $3N \times 3N$ matrix $\boldsymbol{\mathcal{R}}$ is the configuration-space version of the symmetry operator r, and $\boldsymbol{\Gamma}^n(r)$ denotes the corresponding irreducible representation matrix, of dimensionality l_n. Owing to the inversion symmetry in C_{60}, the displacement patterns are either invariant under inversion (even modes), or change sign (odd modes). There are 10 irreducible representations of I_h. By reducing the 180-dimensional representation generated by transforming the basis formed from three orthogonal unit vectors on each atom, one can *a priori* predict the mode symmetry types. It is found that the number of linearly independent times each irreducible representation appears is $2(A_g)$, $1(A_u)$, $4(T_{1g})$, $5(T_{1u})$, $4(T_{3g})$, $5(T_{3u})$, $6(G_g)$, $6(G_u)$, $8(H_g)$, and $7(H_u)$. The notation is that of *Weeks* and *Harter* [2.16], who provide a detailed account of the icosahedral group and its application to the harmonic vibrations of C_{60}. The irreducible representations of types A, T, G, and H are of dimensionalities 1, 3, 4, and 5, respectively, these giving the degeneracies, while the g and u subscripts specify the representations for even and odd modes. The six zero-frequency modes corresponding to rigid translations and rotations account for one appearance each of T_{1u} and T_{1g}, respectively, and when account is taken of the dimensionalities of the remaining irreducible representations, it is seen that there are just 46 distinct frequencies among the 174 vibrational modes.

The high symmetry of a C_{60} molecule also severely restricts the number of spectroscopically active modes in ir absorption and Raman scattering, particularly in first-order processes, i.e., those involving a single vibrational quantum. For absorption, the coupling is through the dipole moment vibrationally induced in the system. In particular, within the adiabatic and dipole approximations, the interaction between a molecule and an external electric field is $-\vec{\boldsymbol{\mu}}(\boldsymbol{u}) \cdot \vec{\boldsymbol{E}} \exp(-\mathrm{i}\omega t)$, where $\vec{\boldsymbol{\mu}}(\boldsymbol{u})$ is the expectation value of the total dipole moment in the molecule for a fixed atomic configuration $\boldsymbol{u}$:

$$\vec{\boldsymbol{\mu}}(\boldsymbol{u}) = \sum_{\ell} q_\ell \vec{\boldsymbol{u}}(\ell) + \int d^3x\, \phi_0^*(\boldsymbol{x}, \boldsymbol{u}) \sum_{e} q_e \vec{\boldsymbol{x}}_e\, \phi_0(\boldsymbol{x}, \boldsymbol{u}). \tag{2.13}$$

The arrows denote 3-D Cartesian vectors, q_ℓ is the nuclear charge on atom ℓ, the sums are over all of the atoms and electrons, and the expectation value is taken with respect to the electronic coordinates. Physically, the second term describes the electronic charge redistribution induced by atomic motion. Expanding $\vec{\boldsymbol{\mu}}(\boldsymbol{u})$ in powers of the atomic displacements and using (2.8) to express the result in terms of the normal coordinates gives

$$\vec{\boldsymbol{\mu}}(\boldsymbol{u}) = \sum_{f} \vec{\boldsymbol{\mu}}_f d_f + \frac{1}{2} \sum_{ff'} \vec{\boldsymbol{\mu}}_{ff'}\, d_f d_{f'} + \cdots, \tag{2.14}$$

where it is assumed that the system has no permanent dipole moment. The first two coefficients are given by

$$\vec{\boldsymbol{\mu}}_f = \sum_{\ell\alpha} \left[\frac{\partial \vec{\boldsymbol{\mu}}(\boldsymbol{u})}{\partial u(\ell\alpha)} \right]_0 \chi(\ell\alpha|f) \tag{2.15}$$

and

$$\vec{\boldsymbol{\mu}}_{ff'} = \sum_{\ell\alpha, \ell'\alpha'} \left[\frac{\partial^2 \vec{\boldsymbol{\mu}}(\boldsymbol{u})}{\partial u(\ell\alpha) \partial u(\ell'\alpha')} \right]_0 \chi(\ell\alpha|f)\, \chi(\ell'\alpha'|f'). \tag{2.16}$$

Within standard linear response theory and the harmonic approximation, the terms in (2.14) lead to absorption involving one, two, ... vibrational quanta, respectively. Specifically, for a sample containing C molecules per unit volume, the first-order contribution to the absorption is given by

$$A(\omega) = \frac{2\pi^2 C}{3cn} \sum_f |\vec{\boldsymbol{\mu}}_f|^2 \delta(\omega - \omega_f), \tag{2.17}$$

where n is the medium's refractive index, and an orientational average has been performed. Local field effects have not been included. This harmonic approximation result is independent of the temperature. In practice, the delta function absorption lines are broadened phenomenologically, e.g., as Lorentzians. The integrated absorption strength of a mode f is given by the square of the dipole moment derivative $\vec{\boldsymbol{\mu}}_f$, which according to (2.14) is the mode's effective charge.

Again within the harmonic approximation, the second-order dipole moment coefficients $\vec{\boldsymbol{\mu}}_{ff'}$ give rise to temperature-dependent sum and difference absorption involving two modes. Sum and difference absorption can also result from the linear term in (2.14), provided that cubic anharmonicity is present; in this case a virtually excited ir active mode decays anharmonically into the two final state vibrational quanta. Reference [2.17] designates these two mechanisms for two-quantum absorption as "electrical" and "mechanical," respectively.

The ir selection rules derive from the transformation properties of the dipole moment operator $\vec{\boldsymbol{\mu}}(\boldsymbol{u})$ under symmetry operations. We now summarize the standard group theoretic argument for first-order absorption. It is useful to write the β-component of (2.15) compactly as

$$\mu_{\beta,f} = \tilde{\boldsymbol{\mu}}_\beta \boldsymbol{\chi}(f), \tag{2.18}$$

where $\boldsymbol{\mu}_\beta \equiv \{\mu_\beta(\ell\alpha) = [\partial\mu_\beta/\partial u(\ell\alpha)]_0\}$ is the $3N \times 1$ vector of real-space effective charges describing the β-component of the dipole moment. It is straightforward to show that if $\boldsymbol{r} = \{r_{\alpha\alpha'}\}$ is the 3×3 Cartesian matrix for a symmetry operation r, the vectors $\boldsymbol{\mu}_\beta$ transform according to

$$\mathcal{R}\boldsymbol{\mu}_\beta = \sum_{\beta'=1}^{3} r_{\beta'\beta}\boldsymbol{\mu}_{\beta'}; \tag{2.19}$$

that is, $\boldsymbol{\mu}_x$, $\boldsymbol{\mu}_y$, and $\boldsymbol{\mu}_z$ are basis functions for the representation $\boldsymbol{r}$. For an arbitrary system, this representation would generally be reducible, and in view of (2.18) and the standard group theory basis function orthogonality theorem [2.18], $\vec{\boldsymbol{\mu}}_f$ will vanish whenever mode f belongs to an irreducible representation not contained in $\boldsymbol{r}$. For the full icosahedral group, $\boldsymbol{r}$ is just the irreducible representation T_{1u}. Hence only T_{1u} modes can be first-order ir active in pure C_{60}.

For two-quantum absorption, the selection rules for either the second-order dipole moment mechanism (2.16) or the first-order dipole moment plus

Table 2.1. Symmetry-allowed second-order combinations for infrared absorption and off-resonance Raman scattering, under the full icosahedral point group (I_h). The ir- and Raman-allowed combinations are denoted by I and R, respectively. Adapted from [2.70]

	Even parity modes					Odd parity modes				
	A_g	T_{1g}	T_{3g}	G_g	H_g	A_u	T_{1u}	T_{3u}	G_u	H_u
A_g	R				R		I			
T_{1g}		R	R	R	R	I	I			I
T_{3g}		R	R	R	R				I	I
G_g		R	R	R	R			I	I	I
H_g	R	R	R	R	R		I	I	I	I
A_u		I				R				R
T_{1u}	I	I			I		R	R	R	R
T_{3u}				I	I		R	R	R	R
G_u			I	I	I		R	R	R	R
H_u		I	I	I	I	R	R	R	R	R

cubic anharmonicity mechanism are the same: two modes f and f' cannot be second-order active unless their direct product representation $\boldsymbol{\Gamma}^f \times \boldsymbol{\Gamma}^{f'}$ includes the representation $\boldsymbol{r}$. For pure C_{60}, the resulting possible combination mode symmetry types are listed in Table 2.1.

For Raman scattering, we focus on the off-resonance case, when the incident visible photon energy $\hbar\omega_\mathrm{L}$ is well below that of any electronic transitions. The relevant quantity is then the system's electronic polarizability $\mathcal{P}_{\alpha\beta}(\omega_\mathrm{L}, \boldsymbol{u})$ for fixed atomic configuration $\boldsymbol{u}$, given by

$$\mathcal{P}_{\alpha\beta}(\omega_\mathrm{L}, \boldsymbol{u}) = \hbar^{-1} \sum_n \left[\frac{D_\alpha^{0n}(\boldsymbol{u}) D_\beta^{n0}(\boldsymbol{u})}{\omega_{n0}(\boldsymbol{u}) - \omega_\mathrm{L}} + \frac{D_\beta^{0n}(\boldsymbol{u}) D_\alpha^{n0}(\boldsymbol{u})}{\omega_{n0}(\boldsymbol{u}) + \omega_\mathrm{L}} \right] . \tag{2.20}$$

Here, the electronic transition dipole matrix elements are

$$\vec{\boldsymbol{D}}^{0n}(\boldsymbol{u}) = \int d^3x \, \phi_0^*(\boldsymbol{x}, \boldsymbol{u}) \sum_e q_e \vec{\boldsymbol{x}}_e \, \phi_n(\boldsymbol{x}, \boldsymbol{u}), \tag{2.21}$$

and $\omega_{n0}(\boldsymbol{u})$ denotes the frequency of electronic transitions $0 \to n$, for fixed $\boldsymbol{u}$. Note that $\mathcal{P}_{\alpha\beta}(\omega_\mathrm{L}, \boldsymbol{u}) = \mathcal{P}_{\beta\alpha}^*(\omega_\mathrm{L}, \boldsymbol{u})$, so that the real and imaginary parts of this quantity are, respectively, symmetric and antisymmetric in $\alpha\beta$. For scattered light frequency ω_S, the photon differential scattering cross-section is then [2.19]

$$\begin{aligned} \left(\frac{\mathrm{d}^2\sigma}{\mathrm{d}\Omega \mathrm{d}\omega} \right)_{\eta'\eta} &= \frac{\omega_\mathrm{L} \omega_\mathrm{S}^3}{2\pi c^4} \sum_{\alpha\beta\gamma\lambda} \eta'_\alpha \eta_\beta \eta'_\gamma \eta_\lambda \int_{-\infty}^{\infty} \mathrm{d}t \, \exp(-\mathrm{i}t\omega) \\ &\quad \times \langle \mathcal{P}_{\alpha\beta}^*(\omega_\mathrm{L}, \boldsymbol{u}; t) \mathcal{P}_{\gamma\lambda}(\omega_\mathrm{L}, \boldsymbol{u}) \rangle , \end{aligned} \tag{2.22}$$

where $\omega \equiv \omega_{\rm S} - \omega_{\rm L}$, and η' and η are the polarization directions of the scattered and incident light, respectively. The brackets denote a thermal average, and $\mathcal{P}^*_{\alpha\beta}(\omega_{\rm L}, \boldsymbol{u}; t) \equiv \exp(\mathrm{i}tH/\hbar)\mathcal{P}^*_{\alpha\beta}(\omega_{\rm L}, \boldsymbol{u})\exp(-\mathrm{i}tH/\hbar)$, where H is the vibrational Hamiltonian. Similar to the discussion below (2.14), expansion of $\mathcal{P}_{\alpha\beta}(\omega_{\rm L}, \boldsymbol{u})$ in the atomic displacements and the use of (2.8) give

$$\mathcal{P}_{\alpha\beta}(\omega_{\rm L}, \boldsymbol{u}) = \sum_f P_{\alpha\beta,f} d_f + \frac{1}{2}\sum_{ff'} P_{\alpha\beta,ff'}\, d_f d_{f'} + \cdots . \tag{2.23}$$

The first-order coefficients are

$$P_{\alpha\beta,f} = \sum_{\ell\gamma} P_{\alpha\beta}(\ell\gamma)\chi(\ell\gamma|f) = \tilde{\boldsymbol{P}}_{\alpha\beta}\boldsymbol{\chi}(f), \tag{2.24}$$

where the $3N \times 1$ vector $\boldsymbol{P}_{\alpha\beta} \equiv \{P_{\alpha\beta}(\ell\gamma) = [\partial\mathcal{P}_{\alpha\beta}(\omega_{\rm L}, \boldsymbol{u})/\partial u(\ell\gamma)]_0\}$ contains the real-space electronic polarizability derivatives. Within the harmonic approximation, the use of the linear term of (2.23) in (2.22) yields the cross-section for first-order Stokes scattering at frequency shift ω as

$$\left(\frac{\mathrm{d}^2\sigma}{\mathrm{d}\Omega \mathrm{d}\omega}\right)^{1,\text{Stokes}}_{\eta'\eta} = \frac{\hbar\omega_{\rm L}}{2c^4}\sum_{f=1}^{3N}\frac{(\omega_{\rm L}-\omega_f)^3(\langle n(\omega_f)\rangle + 1)}{\omega_f} \times \left|\sum_{\alpha\beta}\eta'_\alpha\eta_\beta P_{\alpha\beta,f}\right|^2 \delta(\omega+\omega_f), \tag{2.25}$$

where $\langle n(\omega_f)\rangle = [\exp(\hbar\omega_f/k_{\rm B}T) - 1]^{-1}$ is the thermal average occupation number of mode f at temperature T. This result is for a single molecule and includes no media or local field corrections. Although the frequency shift of the light is negative for Stokes scattering, we will follow the standard convention and refer to $|\omega|$ as the Raman shift.

The transformation properties of the real-space electronic polarizability derivatives $\boldsymbol{P}_{\alpha\beta}$ under symmetry operations yield the Raman selection rules. The presence of two Cartesian indices leads to a generalization of the earlier dipole moment derivative transformation property (2.19), namely

$$\boldsymbol{\mathcal{R}}\boldsymbol{P}_{\alpha\beta} = \sum_{\alpha'\beta'}(\boldsymbol{r}\times\boldsymbol{r})_{\alpha'\beta',\alpha\beta}\boldsymbol{P}_{\alpha'\beta'} . \tag{2.26}$$

Here $(\boldsymbol{r}\times\boldsymbol{r})_{\alpha'\beta',\alpha\beta} \equiv r_{\alpha'\alpha}r_{\beta'\beta}$ is an element of the 9-dimensional direct product representation $\boldsymbol{r}\times\boldsymbol{r}$. For the full (holohedral) icosahedral group, this is the reducible representation $T_{1u}\times\, T_{1u}$, and it decomposes into the irreducible representations A_g, T_{1g}, and H_g. Again invoking the standard group theory orthogonality theorem for basis functions, we see that the right-hand side of (2.24) vanishes whenever f does not belong to one of these irreducible representations.

One can go further by decomposing the direct product representation $\boldsymbol{r}\times\boldsymbol{r}$ into its symmetric and antisymmetric components, the elements of which are given by $(\boldsymbol{r}\times\boldsymbol{r})^{\mathrm{s,a}}_{\alpha'\beta',\alpha\beta}\equiv r_{\alpha'\alpha}r_{\beta'\beta}\pm r_{\beta'\alpha}r_{\alpha'\beta}$, with the plus sign referring to the symmetric component. For $(\boldsymbol{r}\times\boldsymbol{r})^{\mathrm{s}}$, the indices are restricted according to $\alpha'\leq\beta'$ and $\alpha\leq\beta$, and likewise for $(\boldsymbol{r}\times\boldsymbol{r})^{\mathrm{a}}$, except that the equality is removed. Thus the symmetric and antisymmetric components of $(\boldsymbol{r}\times\boldsymbol{r})$ are 6-dimensional and 3-dimensional, respectively. It is straightforward to show that the combinations $\{\boldsymbol{P}_{\alpha\beta}\pm\boldsymbol{P}_{\beta\alpha}\}$ are basis functions for $(\boldsymbol{r}\times\boldsymbol{r})^{\mathrm{s,a}}$. Since the real and imaginary parts of $\boldsymbol{P}_{\alpha\beta}$ are symmetric and antisymmetric in $\alpha\beta$, respectively, the real part is a basis function for $(\boldsymbol{r}\times\boldsymbol{r})^{\mathrm{s}}$, whereas the imaginary part is a basis function for $(\boldsymbol{r}\times\boldsymbol{r})^{\mathrm{a}}$. For the full icosahedral group, the symmetric component of $T_{1u}\times T_{1u}$ contains A_g and H_g, whereas the antisymmetric component is T_{1g}. For off-resonance Raman scattering, it is customary to use the static limit $\omega_{\mathrm{L}}=0$ in the electronic polarizability, in which case (2.20) shows that the antisymmetric part of the polarizability vanishes identically. Then only the A_g and H_g modes of C_{60} may be first-order Raman active. For $\omega_{\mathrm{L}}\neq 0$, (2.20) again leads to this result provided the electronic eigenfunctions are purely real. On the other hand, if complex eigenfunctions cannot be ruled out and $\omega_{\mathrm{L}}\neq 0$, the three T_{1g} modes are, in principle, first-order allowed in off-resonance Raman scattering.

Equation (2.25) applies to a molecule with a fixed orientation; the indices $\alpha\beta$ refer to axes fixed in the scatterer. It is straightforward to determine the forms of the polarizability derivatives $\{P_{\alpha\beta,f}\rightarrow P_{\alpha\beta,(n,i)}\}$ for the first-order Raman active modes, where the mode index f has again been replaced by (n,i), with n labeling the irreducible representation and i the partner. This is conveniently done using projection operator techniques, by determining the forms of the ℓ_n matrices $\sum_r \Gamma^n_{ki}(r)\tilde{\boldsymbol{r}}\boldsymbol{A}\boldsymbol{r}$ for $i=1,\ldots,\ell_n$ and k held fixed, where $\boldsymbol{A}$ is an arbitrary 3×3 trial matrix and the sum is over all symmetry operations. The results for I_h are given in Table 2.2. For randomly oriented molecules, i.e., C_{60} in solution, one should average over all orientations. For the common cases of 0°, 90°, and 180° scattering geometries, this leads to the standard results in terms of Placzek invariants [2.20]. These quantities are $G^0=\frac{1}{3}|P_{xx}+P_{yy}+P_{zz}|^2$, $G^{\mathrm{s}}=\frac{1}{3}[|P_{xx}-P_{yy}|^2+|P_{xx}-P_{zz}|^2+|P_{yy}-P_{zz}|^2]+\frac{1}{2}[|P_{xy}+P_{yx}|^2+|P_{xz}+P_{zx}|^2+|P_{yz}+P_{zy}|^2]$, and $G^{\mathrm{a}}=\frac{1}{2}[|P_{xy}-P_{yx}|^2+|P_{xz}-P_{zx}|^2+|P_{yz}-P_{zy}|^2]$. If $I_\perp$ and $I_\parallel$ are the scattered intensities with polarization perpendicular and parallel to that of the incident light, the depolarization ratio is $\rho=I_\perp/I_\parallel=(3G^{\mathrm{s}}+5G^{\mathrm{a}})/(10G^0+4G^{\mathrm{s}})$, and the total scattering is $I_{\mathrm{total}}=I_\parallel+I_\perp\propto 10G^0+7G^{\mathrm{s}}+5G^{\mathrm{a}}$. From the results of Table 2.2, the depolarization ratios for first-order scattering by A_g and H_g modes are 0 and 3/4, respectively, while if T_{1g} scattering occurs, it is predicted to have 100% perpendicular polarization. In the remainder of this chapter, we will use selection rules appropriate to the static electronic polarizability, which is real and symmetric, so that only the A_g and H_g modes are first-order Raman active, as noted earlier.

The selection rules for the second-order sum and difference scattering arising from the second term in (2.23) within the harmonic approximation

Table 2.2. Forms of the polarizability derivative tensors for first-order Raman active modes under icosahedral symmetry. The T_{1g} contributions vanish in either the static limit $\omega_{\mathrm{L}} = 0$ or whenever the electronic eigenfunctions are purely real. These forms were obtained via the projection operator technique described above, using the irreducible representation matrices given by *Weeks* and *Harter* [2.16]. The z-axis is a 5-fold axis, and the yz plane is a reflection plane

A_g	$\begin{pmatrix} a & 0 & 0 \\ 0 & a & 0 \\ 0 & 0 & a \end{pmatrix}$				
H_g	$\begin{pmatrix} b & 0 & 0 \\ 0 & b & 0 \\ 0 & 0 & -2b \end{pmatrix}$	$\begin{pmatrix} \sqrt{3}b & 0 & 0 \\ 0 & -\sqrt{3}b & 0 \\ 0 & 0 & 0 \end{pmatrix}$	$\begin{pmatrix} 0 & \sqrt{3}b & 0 \\ \sqrt{3}b & 0 & 0 \\ 0 & 0 & 0 \end{pmatrix}$	$\begin{pmatrix} 0 & 0 & 0 \\ 0 & 0 & \sqrt{3}b \\ \sqrt{3}b & 0 & 0 \end{pmatrix}$	$\begin{pmatrix} 0 & 0 & -\sqrt{3}b \\ 0 & 0 & 0 \\ -\sqrt{3}b & 0 & 0 \end{pmatrix}$
Partner	1	2	3	4	5
T_{1g}	$\begin{pmatrix} 0 & c & 0 \\ -c & 0 & 0 \\ 0 & 0 & 0 \end{pmatrix}$	$\begin{pmatrix} 0 & 0 & 0 \\ 0 & 0 & c \\ -c & 0 & 0 \end{pmatrix}$	$\begin{pmatrix} 0 & 0 & -c \\ 0 & 0 & 0 \\ c & 0 & 0 \end{pmatrix}$		
Partner	1	2	3		

are a straightforward generalization: two modes f and f' cannot be second-order active unless their direct product representation $\boldsymbol{\Gamma}^f \times \boldsymbol{\Gamma}^{f'}$ includes irreducible representations contained in $(\boldsymbol{r} \times \boldsymbol{r})^{\mathrm{s}}$. Thus for full icosahedral symmetry, $\boldsymbol{\Gamma}^f \times \boldsymbol{\Gamma}^{f'}$ must contain A_g and H_g. The resulting allowed second-order mode symmetry types are listed in Table 2.1.

For first-order ir and Raman activity, the net result for pure C_{60} is that there are very few active modes: the ir-active T_{1u} modes (four distinct frequencies), and the Raman-active A_g (two distinct frequencies) and H_g modes (eight distinct frequencies). To investigate the remaining $46 - 14 = 32$ distinct frequencies of the first-order "silent" ir and Raman modes, one can study the much more complicated second-order spectra or introduce perturbations that break the first-order selection rules. The latter approach will, of course, generally change the dynamics. In the case of C_{60}, however, isotopic substitutions and intermolecular interactions provide "gentle" perturbations which can nevertheless be probed by the sensitive first-order optical spectroscopies to yield information on the unperturbed silent modes [2.21, 2.22].

2.1.3 Symmetry-Lowering Perturbations

The theoretical basis of Sect. 2.1.1 applies to both molecules and solids, provided that the Born-Oppenheimer approximation holds, whereas the symmetry considerations of Sect. 2.1.2 are for isolated molecules. This means that the calculated frequencies, eigenvectors, and spectroscopic intensities should be compared with experiments on isotopically pure molecules in the gas phase. However, all of the existing data on the vibrational properties of C_{60} stem from measurements carried out on thin films, crystals, and solutions. Fortunately, the forces between C_{60} molecules in the condensed phases are much weaker than the strong interatomic covalent forces within individual molecules. For instance, at room temperature solid C_{60} is a Van der Waals bound fcc lattice of approximately freely rotating molecules. The weakness of the intermolecular binding is further exemplified by the fact that the highest measured intermolecular phonon frequency at room temperature [2.23] is only $\sim 50\,\mathrm{cm}^{-1}$, well below that of the lowest intramolecular frequency, namely $272\,\mathrm{cm}^{-1}$ for the $H_g(1)$ "squashing" mode.[1] When the temperature is lowered below 260 K, the approximately free rotations are lost, but the intermolecular interactions and their effects on the dynamics remain relatively weak.

Another perturbation that is usually present in experiments on C_{60} arises from the naturally occurring distribution of carbon isotopes in the samples. Since the fractional mass change arising from the substitution of a single ^{12}C atom by a ^{13}C in a C_{60} molecule is 1/720, the dynamical effects induced by isotopic perturbations are also weak.

[1] We are labeling the modes according to their frequencies. Thus the 8 distinct H_g mode frequencies will be listed as $H_g(1)$–$H_g(8)$, each of which is 5-fold degenerate.

In the limit where the perturbations of the icosahedral symmetry are small and the associated changes of the dynamics can be neglected, the spectroscopic observation of silent modes provides direct information on the vibrational frequencies in icosahedral C_{60}, even if the precise form of the perturbation is not known. The intensity of the silent modes will of course be weak in this limit, but the high sensitivity of first-order techniques such as infrared absorption and Raman spectroscopy makes it possible to detect signals which are two or three orders of magnitude weaker than those arising from the strongest optically active modes. Additional information on the vibrational structure of the icosahedral molecule can be obtained when the perturbation is known in detail, allowing one to predict frequency shifts, selection rules, and spectroscopic intensities. Accordingly, we devote the remainder of this section to a discussion of isotopic and intermolecular interaction perturbations, which are not only well-characterized, but in practice are the most significant deviations from icosahedral symmetry. Not covered are extrinsic effects such as the presence of residual amounts of impurities, solvents, or other fullerenes, although the experimentalist must always keep these possibilities in mind. For example, it is known that even minute amounts of solvents can change the crystalline structure [2.24], and this is likely to affect the vibrational properties.

Isotopes

The simplest perturbation of the ideal icosahedral molecule arises from the presence of different carbon isotopes. The natural abundance of ^{13}C is 1.1%, with the result that 49% of the molecules obtained from natural graphite contain one or more ^{13}C isotopes. Studies of the isotopic distribution within the molecule show that it is truly random [2.25]. Excluding the unlikely possibility of significant zero-point effects, the mass perturbation does not affect the electronic structure of the molecule, so that force constants are unchanged. Hence calculations of the normal modes of icosahedral C_{60} can easily be extended to include the case of isotopic disorder, with no loss of accuracy. Details of such calculations are given in [2.21, 2.22, 2.26], and are summarized here in Sect. 2.2.5. These studies show that the analysis of the frequency shifts of modes which are spectroscopically active for the unperturbed molecule provides information on silent modes. Furthermore, the invariance of the electronic structure under isotopic substitutions means that the IR and Raman intensities in isotopically disordered C_{60} can be computed using the same effective charges and electronic polarizability derivatives that are used for icosahedral C_{60}. For random isotopic disorder, all modes should become Raman and IR active, and their intensity can be predicted with accuracy.

Solid-State Effects

The second form of commonly occurring symmetry lowering arises from the interactions between C_{60} molecules. This occurs not only in solid phases but also in solution, where C_{60} molecules have been shown to form aggregates [2.27]. The intermolecular interactions in crystalline C_{60} have been studied in much more detail, although a detailed analysis of the latest X-ray diffuse scattering experiments suggests that none of the models proposed so far is entirely satisfactory [2.28]. In order to understand the new spectroscopic features associated with the formation of C_{60} crystals, we must recast our discussion of the C_{60} vibrations using the standard theoretical framework for a periodic system. As mentioned above, the C_{60} molecules occupy fcc sites in the solid phase [2.29], and at room temperature, the rapid rotational diffusion of each molecule leads to an effective fcc crystal structure. At $T \approx 260\,\mathrm{K}$ the crystal undergoes a phase transition, below which the four molecules in the conventional fcc cubic cell remain in the fcc sites but become orientationally inequivalent [2.23, 2.24, 2.30]. The orientational alignment occurs over a large temperature range, and the resulting low-temperature structure is simple cubic with four molecules per unit cell. The space group is $P21/a$ (short $Pa\bar{3}$) [2.31].

In a lattice, the equilibrium site index ℓ of Sect. 2.1.1 is split: $\ell \to (\ell, b)$, where ℓ now labels the primitive unit cell of the Bravais lattice and the sites within each cell are labeled by the basis index b; thus for crystalline C_{60}, b ranges from 1 to $60s$, where s is the number of inequivalent molecules in each unit cell. The elements of the harmonic force constant matrix of (2.3) become $\Phi(\ell b\alpha, \ell' b'\alpha')$, and lattice periodicity renders them invariant to the addition of any cell index ℓ'' to both ℓ and ℓ'. Bloch's theorem then results in the normal modes being plane waves

$$\chi(\ell b\alpha|\vec{k}j) = \frac{e(b\alpha|\vec{k}j)}{\sqrt{N_c m_b}} \exp[\mathrm{i}\vec{k} \cdot \vec{R}(\ell b)], \tag{2.27}$$

where $\vec{R}(\ell b)$ is the equilibrium position of site (ℓb), and periodic boundary conditions have been applied over a supercell containing N_c unit cells. The mode label f of Sect. 2.1.1 is now split into a wave vector $\vec{k}$ and polarization index j, with $j = 1, \ldots, 180s$. The $180s$-dimensional phonon polarization vectors $\boldsymbol{e}(\vec{k}j) \equiv \{e(b\alpha|\vec{k}j)\}$ are eigenvectors of the dynamical matrix, which is given by the lattice Fourier transform of the force constant matrix:

$$D(b\alpha, b'\alpha'|\vec{k}) = \sum_{\ell'} \frac{\Phi(\ell b\alpha; \ell' b'\alpha')}{\sqrt{m_b m_{b'}}} \exp\{-\mathrm{i}\vec{k} \cdot [\vec{R}(\ell b) - \vec{R}(\ell' b')]\}. \tag{2.28}$$

Finally, without loss of generality the periodic boundary condition supercell can be taken with an equal number L of primitive unit cells along each of the 3 Bravais lattice basis vectors, in which case $N_c = L^3$ and the allowed $\vec{k}$'s are given by $\vec{k} = \sum_{i=1}^{3} h_i \vec{b}_i / L_i$, where the three $\vec{b}_i$ are the basis vectors of the reciprocal lattice and the h_i's are positive or negative integers. The $\vec{k}$'s are restricted to the first Brillouin zone and total N_c in number.

In Sect. 2.1.2 we discussed the spectroscopic selection rules for molecular modes with index f. In the crystalline phase, periodicity imposes selection rules involving the wave vector $\vec{k}$. For either ir or Raman spectroscopy, wave vector conservation plus the fact that the photon wavelengths are essentially infinite on the scale of the intermolecular spacing, require that the sum of the wave vectors of the excited and/or destroyed modes must be zero. Thus for first-order absorption or scattering, only $\vec{k} = 0$ modes are active, while in second order one must have $\vec{k}_1 + \vec{k}_2 = 0$. Additional selection rules are obtained by considering the symmetry properties of the group of the wave vector associated with the phonons. For first-order spectroscopies, the group of $\vec{k} = 0$ is needed, and for solid C_{60} this is T_h, the point group of the lattice's space group. Second-order processes involve pairs of phonons with individual wave vectors away from $\vec{k} = 0$. Then the corresponding wave vector groups have lower symmetry, usually resulting in many more allowed combinations.

As noted, C_{60} forms a molecular solid for which the intermolecular interactions are much weaker than the intramolecular covalent forces. In such solids it is simple to track the evolution of the vibrational modes from isolated molecules to crystals. For C_{60}, the phonon dispersion relations $\omega(\vec{k}, j)$ decouple into a series of optic branches corresponding to the intramolecular vibrations above $\sim 270\,\text{cm}^{-1}$ and a second set of lower frequency branches which involve the molecules vibrating essentially as whole entities, i.e., as rigid balls, both translationally and rotationally (librations). The dispersion relations for the high frequency intramolecular optic branches reflect the weak ball-ball interactions and are therefore quite flat.

Well above $T = 260\,\text{K}$, the orientational restoring forces between the balls can be neglected and there are no librational branches. Thus one can describe the low-frequency intermolecular modes to an excellent approximation as those of a simple monatomic fcc lattice of spherical "atoms," each with mass 12×60 amu, giving rise to three acoustic phonon branches.

In the low temperature simple cubic phase, the entire vibrational structure can be understood qualitatively by noting that the orientational forces which cause the phase transition are even weaker than the binding forces responsible for the average fcc structure. As a result, one can obtain a zero-order approximation to the vibrations in the low temperature phase by simply "folding" the fcc Brillouin zone onto the smaller Brillouin zone of the simple cubic lattice with four molecules per unit cell. The weak reorientational potential induces small mode shifts and generally splits the degeneracies arising from the folding process. The folding of the three acoustic branches of the fcc lattice yields 12 low-frequency intermolecular phonon branches, of which 3 are acoustic and 9 are optical. In addition, the three rotational degrees of freedom of each molecule lead to 12 branches corresponding to librational motions. The highest observed librational frequency is of order $30\,\text{cm}^{-1}$ [2.23]. This is lower than the highest translational frequency of $50\,\mathrm{cm}^{-1}$ and confirms the relative weakness of the reorientational potential. Finally, the folding of the 174 high-frequency fcc intramolecular optic branches derived from the modes of the isolated molecules yields 4×174 intramolecular optic branches, which are again relatively flat.

Table 2.3 gives group-theoretic predicted splittings of the $\vec{k} = 0$ intramolecular modes in the solid phase. The first two columns give the splittings for a hypothetical ordered fcc structure in which the crystal's point symmetry about the origin of each molecule is T_h. This is achieved by orienting each C_{60} molecule with the 2-fold axes (which bisect the hexagon-hexagon bonds) parallel to the three orthogonal Cartesian axes. This is the so-called "standard structure" [2.32], and the space group is $Fm\bar{3}$. It should be noted that there are two equivalent standard structures, which are related by a reflection in the (110) planes. In real crystals, this leads to merohedral twinning and disorder, as will be discussed below. For T_h symmetry, modes belonging to the A_g, E_g, and T_g irreducible representations are first-order Raman-active, whereas the T_u modes are first-order ir-active. The third column in Table 2.3 is for reference and represents the situation for a hypothetical sc structure with four non-interacting I_h C_{60} molecules per unit cell. The last column gives the predicted splittings for the actual low temperature simple cubic phase, of space group $Pa\bar{3}$. In this structure, the four molecules in the simple cubic unit cell are again centered at fcc sites, but they are each rotated from their standard-structure orientation by the same angle, about different $\langle 111 \rangle$ axes. Although the point symmetry at each C_{60} site is then S_6, the presence of non-symmorphic operations (screw axes and glide planes) in $Pa\bar{3}$ results in the relevant point group for $\vec{k} = 0$ once again being T_h.[2] Hence the symmetry properties of this group still provide the spectroscopic selection rules for the first-order Raman- and ir-active phonons. Note from the second column of Table 2.3 that the formation of the hypothetical standard-structure fcc solid would split only the 4-fold and 5-fold degenerate modes of I_h C_{60}, whereas the last column shows that all of the levels are split by the reorientational potential in the low-temperature sc phase. Accordingly, one would generally expect the icosahedral A and T mode splittings, which arise only from the reorientational potential, to be smaller than the G and

[2] For a mode at $\vec{k} = 0$, the displacement patterns are the same in each unit cell and are therefore invariant under any translation of the Bravais lattice. Hence the crystal symmetry operations relevant for $\vec{k} = 0$ modes are those involving no Bravais translations, namely pure point symmetry operations and nonsymmorphic operations consisting of point operations (rotation or reflection) followed by a fractional primitive cell translation (see [2.33]). These operations form a group which is isomorphic to one of the 32 crystallographic point groups, and for the space group $Pa\bar{3}$ this is T_h. Thus the group of the wave vector for the low temperature simple cubic phase of C_{60} is T_h, despite the fact that the highest site symmetry in this space group is S_6, occurring at each of the four C_{60} sites.

Table 2.3. Compatibility relations between the I_h C_{60} mode symmetries ($\omega \neq 0$) and the $\vec{k} = 0$ mode symmetries for the fcc standard-configuration lattice (space group $Fm\bar{3}$, point group T_h) and the simple cubic lattice (space group $Pa\bar{3}$, point group T_h, 4 molecules/cell). Modes belonging to A_g, E_g, and T_g are first-order Raman-active, whereas the T_u modes are first-order ir-active. Adapted from [2.11]

I_h	fcc ($Fm\bar{3}$)	I_h	sc ($Pa\bar{3}$)
A_g	A_g	$4A_g$	$A_g + T_g$
T_{1g}	T_g	$4T_{1g}$	$A_g + E_g + 3T_g$
T_{3g}	T_g	$4T_{3g}$	$A_g + E_g + 3T_g$
G_g	$A_g + T_g$	$4G_g$	$2A_g + E_g + 4T_g$
H_g	$E_g + T_g$	$4H_g$	$A_g + 2E_g + 5T_g$
A_u	A_u	$4A_u$	$A_u + T_u$
T_{1u}	T_u	$4T_{1u}$	$A_u + E_u + 3T_u$
T_{3u}	T_u	$4T_{3u}$	$A_u + E_u + 3T_u$
G_u	$A_u + T_u$	$4G_u$	$2A_u + E_u + 4T_u$
H_u	$E_u + T_u$	$4H_u$	$A_u + 2E_u + 5T_u$

H mode splittings, which arise from both the formation of the solid and the reorientational potential.

A significant feature of the fully ordered crystal is that the T_h point group preserves the inversion symmetry, so that the first-order Raman and ir activities of the modes remain mutually exclusive, as for the ideal icosahedral molecule. This could be used to decide which symmetry-breaking mechanism is more important: if the isotopic perturbation dominates, it should be possible to observe the same silent modes in Raman and ir experiments. On the other hand, if the crystal field perturbation is stronger, one should see even-parity silent modes in Raman experiments and odd-parity silent modes in ir. Unfortunately, this comparison cannot be made easily because the weakness of the orientational forces prevents an ideal fully-ordered crystalline structure from being realized experimentally.

The most obvious deviation from an ordered crystalline structure occurs in the room temperature solid phase, where the molecules are rapidly changing their orientation. It was assumed originally that the C_{60} molecules were essentially free rotors about their lattice sites. This assumption was supported by a considerable amount of experimental data, including the excellent fit of the X-ray diffraction pattern by assuming the molecule to be a spherical shell [2.24, 2.30]. The effect of rotations on the spectroscopic activity of the molecule depends on whether the vibrations "see" an averaged external perturbation or a frozen configuration. In the first case, the observed spectra should be similar to those expected from C_{60} molecules in the gas phase. In the latter case, one should observe mode splittings and silent mode activation due to a virtually static random perturbation produced by the instantaneous orientation of the neighboring molecules. According to *Martin* et al. [2.34], the correct limit can be decided by comparing the lifetime (not the period) of the vibration relative to the rotational diffusion time. Applying this criterion to the T_{1u} modes, *Martin* et al. estimate an upper limit of 60 ps for their lifetime [2.34]. On the other hand, from the rotational diffusion constant measured with neutron scattering they estimate that a 60° rotation takes about 130 ps, so that it appears that crystalline C_{60} is closer to the frozen configuration limit [2.34]. More recent neutron scattering and X-ray data present a more complicated picture of the room temperature phase [2.23, 2.35, 2.36]. The data show non-negligible orientational correlations between neighboring molecules. The orientational distribution is found not to be isotropic. An excess atomic density of the order of 10% was found near the $\langle 110 \rangle$ directions, whereas a 16% deficit was found along the $\langle 111 \rangle$ directions [2.35]. This reflects a strong tendency for 5-fold axes to point along $\langle 110 \rangle$ directions. This orientation is nearly nine times more likely than the less favored orientation of the 5-fold axes along $\langle 111 \rangle$ directions [2.23]. The existence of these anisotropic orientations breaks the cubic and inversion symmetry and can lead to the activation of silent modes which are forbidden according to Table 2.3, even if the lifetime criterion discussed by *Martin* et al. [2.34] is not fulfilled.

We now discuss the low-temperature phase of solid C_{60} in more detail. As noted above, the effective high temperature fcc structure undergoes a phase transition at about 260 K, to the simple cubic $Pa\bar{3}$ phase, with four molecules per unit cell. In the fully-ordered version of this phase, the molecules remain centered at the fcc sites $\vec{t}_0 = (0,0,0)$, $\vec{t}_1 = (0,1/2,1/2)$, $\vec{t}_2 = (1/2,0,1/2)$, and $\vec{t}_3 = (1/2,1/2,0)$, but they have different orientations. This can be visualized by starting with the hypothetical fcc "standard structure" discussed above and rotating each molecule by the same angle Γ about a different $\langle 111 \rangle$ axis. The value of Γ is not determined by symmetry, but corresponds to an orientation for which an electron-rich double bond in one molecule faces an electron-poor pentagon face on its neighbor.

There are two sources of disorder in the $Pa\bar{3}$ crystal structure of C_{60} that limit the applicability of the $\vec{k} = 0$ mode splittings given in the last column of Table 2.3. The first type of disorder arises from the fact that there are eight distinct but equivalent $Pa\bar{3}$ structures, arising from the different choices for the four distinct $\langle 111 \rangle$ axes about which the molecules are rotated [2.31]. These eight structures can be conveniently described in terms of just one of the axis sets, as follows. Starting with the system in one of the two standard structures and rotating each molecule by the same angle about the axes $\langle 111 \rangle$, $\langle \bar{1}\bar{1}1 \rangle$, $\langle \bar{1}1\bar{1} \rangle$, $\langle 1\bar{1}\bar{1} \rangle$, at positions (0,0,0), (1/2,0,1/2), (0,1/2,1/2), (1/2,1/2,0), respectively, yields a fully ordered $Pa\bar{3}$ structure. Then applying the translations $\vec{t}_1$, $\vec{t}_2$, and $\vec{t}_3$ yields three additional $Pa\bar{3}$ structures [2.32]. The same procedure applied to the other standard configuration then gives four additional $Pa\bar{3}$ structures. The eight total structures thus naturally divide into two sets, and each set gives a different contribution to the X-ray scattering intensity. The alternative structures coexist in real crystals in the form of macroscopic domains [2.37], yielding the so-called "merohedral twinning" [2.29,2.30]. The expression "merohedral disorder" has also been used in this context, but it is more properly applied to the case of alkali-metal fullerene intercalation compounds, where the two standard orientations become energy minima due to the alkali-metal–C_{60} interactions, and the molecules are randomly found in one of these two orientations [2.24]. The presence of merohedral twinning in crystalline C_{60} may scramble the polarization selection rules for Raman scattering, but is unlikely to activate silent modes other than those allowed by the fully ordered $Pa\bar{3}$ structure. This is because twinning does not destroy the local symmetry, and the interaction between distant C_{60} molecules across domain boundaries should be negligible. However, a second source of disorder in low-temperature C_{60} crystals does alter the local symmetry and may activate silent modes not allowed by the $Pa\bar{3}$ symmetry. The origin of this disorder is the existence of a secondary energy minimum for a rotation about a $\langle 111 \rangle$ axis, which brings a double bond in one molecule close to a hexagonal face in a neighboring molecule (instead of a pentagonal face, as in the ground state) [2.23, 2.24, 2.30, 2.38]. Below the phase transition at 260 K, the molecules librate about their equilibrium positions and occasionally jump between these two energy minima. At about 90 K the jumps are frozen on a laboratory time scale, and the crystal exhibits a residual disorder, whereby 84% of the molecules occupy the double-bond/pentagon configuration and the remaining 16% occupy the double-bond/hexagon configuration [2.38]. The existence of this disorder destroys the cubic and inversion symmetry and may activate all silent modes.

In summary, the symmetry of the ideal low-temperature crystalline C_{60} structure permits Raman scattering by all even parity molecular modes and infrared absorption by all odd parity molecular modes. Perturbations which destroy the inversion symmetry, such as the presence of isotopes or the existence of local orientational disorder due to the nearby energy minima can render all modes Raman and infrared active. Which of these mechanisms dominates can only be answered experimentally. The corresponding evidence is discussed in Sect. 2.2.

2.1.4 Survey of Theoretical Calculations

While spectroscopic work on C_{60} vibrations became possible only after the Krätschmer–Huffman breakthrough in 1990, theoretical studies began to appear as early as 1987 [2.39]. The success of the theoretical predictions can now be evaluated by comparing the calculated frequencies with experimentally determined values, as is done in Table 2.4. However, owing to the relatively small number (14) of frequencies which are directly accessible via the first-order optical spectroscopies, the comparison must be done with care. The frequency assignments in Table 2.4 will be discussed extensively in Sect. 2.2, where it will be shown that even with the addition of inelastic neutron scattering, second-order ir and Raman data, and fluorescence data, only 29 out of the total of 46 distinct vibrational frequencies are known with high certainty. This precludes a definitive evaluation of the accuracy of the different theoretical approaches. Another complication arises because, as pointed out in Sect. 2.1.2, each icosahedral irreducible representation except A_u appears more than once in the symmetry decomposition of the 174 vibrational modes. Comparisons between theory and experiment and among theoretical models themselves are usually made by simply ordering the modes according to their frequencies. However, a different match could be obtained by comparing eigenvectors. For example, the $G_u(2)$ and $G_u(3)$ modes of Table 2.4 are quite close in frequency, and the eigenvector of the lower frequency mode – which the experimentalist would call $G_u(2)$ – might actually be closer to the eigenvector of the G_u mode with the theoretically predicted higher frequency, which would be called $G_u(3)$ by the theorist. Unfortunately, the sheer amount of data necessary to specify each of the 180-dimensional C_{60} mode eigenvectors, their nonuniqueness with respect to mode degeneracies, and the difficulty of obtaining them experimentally have precluded eigenvector comparisons. As far as we know, the only direct experimental studies of C_{60} mode eigenvectors is the recent neutron scattering work of *Heid* and coworkers [2.40].

Conceptually, perhaps the simplest vibrational model applied to C_{60} results from a continuum approximation in which the molecule is treated as a thin elastic spherical shell. While the predictive accuracy of such a model may not be sufficient for spectroscopic assignments, the vibrational analysis of a spherical shell can contribute to the qualitative understanding of the molecular modes, particularly those with lower frequencies. A case in point is the near degeneracy of the $T_{3u}(1)$ and $G_u(1)$ modes of C_{60}. This quasi degeneracy is not accidental if one considers these modes to originate from the 7-fold degenerate $\ell = 3$ multiplet of the $O(3)$ group. This multiplet is split by the icosahedral perturbation, and the smallness of the observed splitting suggests that a continuum model might be a useful approximation for the lowest-energy modes of C_{60}. However, in spite of its conceptual simplicity, the problem of an elastic spherical shell is rather complicated analytically, and the two published papers following this idea disagree in their detailed predictions [2.41, 2.42].

Prior to 1990, normal mode calculations treating the C_{60} molecule as a discrete system used force constants transferred from other molecules containing carbon-carbon bonds, such as benzene [2.16], or relied on semi-empirical quantum chemistry calculations (e.g., MNDO, AM1, or QCFF/PI [2.39, 2.43, 2.44, 2.45]). Within these early calculations, the best agreement for the 29 highlighted frequencies in Table 2.4 holds for the QCFF/PI calculations of *Negri* et al. [2.45], with an average deviation from experiment of 5.7%.

First principles calculations based on density functional theory within the local density approximation (LDA) began to appear after 1990. Such calculations give the best agreement with the known C_{60} vibrational frequencies,

Table 2.4. Recommended values (in cm^{-1}) for the 46 distinct normal mode frequencies of an isolated C_{60} molecule. The symmetry types of the ten first-order Raman and four first-order infrared active modes are labeled with bold-face characters in the first column. Bold type is also used in the second column for the frequencies of modes which have been clearly identified on the basis of spectroscopic selection rules or inelastic neutron scattering data. Whenever possible, low-temperature experimental values were used. The third and fourth columns give representative first-principles LDA-based theoretical results, together with their percent errors, as discussed in the text

Mode	Frequency	Theoretical Calculations *Giannozzi* and *Baroni* [2.47]	*Adams* et al. [2.53]
$\boldsymbol{H_g(1)}$	**272**	259 (−4.8%)	259 (−4.8%)
$T_{3u}(1)$	**342**	337 (−1.5%)	330 (−3.5%)
$G_u(1)$	**353**	349 (−1.1%)	353 (0.0%)
$H_u(1)$	**403**	399 (−1.0%)	399 (−1.0%)
$\boldsymbol{H_g(2)}$	**433**	425 (−1.8%)	427 (−1.4%)
$G_g(1)$	**485**	480 (−1.0%)	484 (−0.2%)
$\boldsymbol{A_g(1)}$	**496**	495 (−0.2%)	494 (−0.4%)
$\boldsymbol{T_{1u}(1)}$	**526**	527 (0.2%)	522 (−0.8%)
$H_u(2)$	**534**	530 (−0.7%)	533 (−0.2%)
$T_{3g}(1)$	**553**	548 (−0.9%)	547 (−1.1%)
$G_g(2)$	**567**	566 (−0.2%)	554 (−2.3%)
$T_{1g}(1)$	**568**	564 (−0.7%)	565 (−0.5%)
$\boldsymbol{T_{1u}(2)}$	**575**	586 (1.9%)	570 (−0.9%)
$H_u(3)$	**668**	662 (−0.9%)	654 (−2.1%)
$\boldsymbol{H_g(3)}$	**709**	711 (0.3%)	694 (−2.1%)
$G_g(3)$	**736**	762 (3.5%)	745 (1.2%)
$H_u(4)$	**743**	741 (−0.3%)	727 (−2.2%)
$T_{3u}(2)$	753	716 (−4.9%)	696 (−7.6%)
$T_{3g}(2)$	756	767 (1.5%)	717 (−5.2%)
$G_u(2)$	764	748 (−2.1%)	708 (−7.3%)
$\boldsymbol{H_g(4)}$	**772**	783 (1.4%)	760 (−1.6%)
$G_u(3)$	776	782 (0.8%)	753 (−3.0%)
$T_{3g}(3)$	796	794 (−0.3%)	757 (−4.9%)
$T_{1g}(2)$	831	823 (−1.0%)	813 (−2.2%)
$G_u(4)$	961	975 (1.5%)	970 (0.9%)
$T_{3u}(3)$	973	993 (2.1%)	954 (−2.0%)
A_u	984	943 (−4.2%)	929 (−5.6%)
$G_g(4)$	1079	1118 (3.6%)	1123 (4.1%)
$\boldsymbol{H_g(5)}$	**1099**	1120 (1.9%)	1103 (0.4%)
$\boldsymbol{T_{1u}(3)}$	**1182**	1218 (3.0%)	1227 (3.8%)
$T_{3u}(4)$	1205	1228 (1.9%)	1239 (2.8%)
$H_u(5)$	**1223**	1231 (0.7%)	1243 (1.6%)
$\boldsymbol{H_g(6)}$	**1252**	1281 (2.3%)	1328 (6.1%)

continued on following page

Table 2.4 continued

Mode	Frequency	Theoretical Calculations *Giannozzi* and *Baroni* [2.47]	*Adams* et al. [2.53]
$T_{1g}(3)$	1289	1296 (0.5%)	1309 (1.6%)
$G_u(5)$	**1309**	1334 (1.9%)	1369 (4.6%)
$G_g(5)$	1310	1322 (0.9%)	1332 (1.7%)
$H_u(6)$	1344	1363 (1.4%)	1387 (3.2%)
$T_{3g}(4)$	1345	1363 (1.3%)	1385 (3.0%)
$G_u(6)$	**1422**	1452 (2.1%)	1525 (7.2%)
$\boldsymbol{H_g(7)}$	**1425**	1450 (1.8%)	1535 (7.7%)
$\boldsymbol{T_{1u}(4)}$	**1429**	1462 (2.3%)	1560 (9.2%)
$\boldsymbol{A_g(2)}$	**1470**	1504 (2.3%)	1607 (9.3%)
$G_g(6)$	1482	1512 (2.0%)	1578 (6.5%)
$T_{3u}(5)$	1525	1535 (0.7%)	1598 (4.8%)
$H_u(7)$	**1567**	1569 (0.1%)	1622 (3.5%)
$\boldsymbol{H_g(8)}$	**1575**	1578 (0.2%)	1628 (3.4%)

with an average deviation ranging from 1.8% to 3.9%. One of the main differences between the various implementations of the LDA approach stems from the type of basis functions used for the electronic states. Expansions in orthogonal plane waves are advantageous in that the convergence of a calculation can be improved by simply adding more plane waves. Moreover, this basis lends itself to the derivation of convenient perturbative expressions for the dynamical matrix of (2.28). This Density Functional Perturbation Theory approach [2.46] was used by *Giannozzi* and *Baroni* [2.47] to compute the vibrational frequencies of C_{60}, with an average deviation from experiment of only 1.8%. These frequencies are given in the second column of Table 2.4. Another well-known plane-wave implementation of LDA techniques is that of *Car* and *Parrinello* [2.48], which uses atomic forces computed directly from the electronic states to perform molecular dynamics calculations. An early application of this method to the dynamics of C_{60} is given in [2.49]. A plane wave basis necessitates the use of periodic boundary conditions, and for isolated molecules the associated supercell is simply taken to be much larger than the molecule.

Since plane waves are not particularly well-suited for describing compact atomic orbitals such as those in fullerenes, plane wave expansions can become computationally very intensive (24000 plane waves were used for the molecular C_{60} calculations of [2.47]). The use of a small number of localized basis functions is a natural choice for such systems. Several methods of this type have been applied to the molecular properties of C_{60} [2.50, 2.51, 2.52]. The agreement with experiment is excellent. *Quong* et al. [2.51] and *Wang* et al. [2.52] used different all-electron localized basis sets within the LDA and obtained average deviations from the experimental frequencies of 1.9% and 2.2%, respectively, very close to the results of *Giannozzi* and *Baroni* [2.47]. In 1991, *Adams* et al. [2.50] used a minimal sp^3 basis of local pseudoatomic orbitals for each atom and obtained an average deviation from experiment of $\approx 10\%$. Subsequent refinements [2.53] have improved this to 3.9%. The general method they employed was developed by *Sankey* et al. [2.54] and utilizes approximations which render it computationally very efficient, giving it the significant advantage of being suitable for much larger fullerenes, including tubes and capsules [2.55] as well as polymerized fullerenes [2.53]. Because this method underlies some of our subsequent discussion, we have included its predicted frequencies in Table 2.4 and will now briefly highlight a few of its key points.

First, the local orbitals are computed from a self-consistent LDA pseudopotential calculation for the free atom, but subject to the boundary condition that the wave functions vanish at a finite cutoff radius r_c, which for fullerenes is taken at 4.1 Bohr radii[3] This has the effect of slightly exciting the orbitals from their ground states, crudely representing the effects of electronic confinement in the molecule. More important from a computational standpoint is the fact that this vastly decreases the number of electronic overlap integrations that must be performed. Second, the Kohn-Sham self-consistent energy functional is replaced by the much simpler non self-consistent *Harris* energy functional [2.56], which takes the input charge density to be a sum of the neutral atom densities. The elimination of the need to compute the electronic charge density self-consistently, together with the finite orbital confinement, results in a computationally very efficient method to obtain the atomic potential energy function $V(\boldsymbol{u}) = E_0(\boldsymbol{u})$ of Sect. 2.1.1 for an arbitrary atomic configuration. Moreover, methods based on the Hellmann-Feynman theorem can be used for accurate calculations of the corresponding atomic forces $F(\ell\alpha) = -\partial V(\boldsymbol{u})/\partial u(\ell\alpha)$ directly from the electronic states [2.57], avoiding the necessity of numerically differentiating V. One can then either obtain the harmonic force constant matrix as discussed below or perform molecular dynamics simulations of real-time atomic motions, as in the Car-Parrinello method. The net result is a computationally fast and entirely real-space scheme to calculate the electronic structure, dynamics, and equilibrium geometry. No *a priori* symmetrization of the electronic states is made. For C_{60}, the calculations reproduce the experimental bond lengths, and yield all of the mode frequencies with an average deviation of 3.9%, as noted above.

[3] Note that this value $r_c = 4.1$ Bohr radii differs from the original value (3.3 Bohr radii) used in [2.50]. As is detailed in the first and third publications cited in [2.53], the use of 4.1, together with a minor alteration in the calculation of the so-called electron double-counting correction, substantially improves the results of [2.50]. The measured bond lengths of C_{60} are reproduced, and the average deviation of the calculated frequencies of the 14 first-order infrared and Raman active modes is 3.7%, compared with $\approx 10\%$ in [2.50].

In computing the dynamics of C_{60} from first-principles, the high symmetry of the molecule greatly simplifies the work. A given row (or column) of the harmonic force constant matrix can be obtained simply by displacing one of the atoms and computing the resulting force per unit displacement on each atom. Specifically, if the atom at ℓ' is given a displacement $u(\ell'\alpha')$ in the direction α' and the remaining atoms are fixed at their equilibrium positions, Equation (2.3) gives $\Phi(\ell\alpha, \ell'\alpha') = -F(\ell\alpha)/u(\ell'\alpha')$. Since all of the equilibrium sites of C_{60} are symmetrically equivalent only one atom must be displaced, and the remaining elements of the force constant matrix can be obtained by applying point symmetry operations. This method was employed in the local orbitals first-principles calculations of [2.51,2.52,2.53]. For example, the frequencies given in the fourth column of Table 2.4 were obtained by first displacing a single atom by 0.0125 Å along $\pm x, \pm y, \pm z$. For each of the six configurations, the force (per unit displacement) was computed at each site, and the results for plus and minus dispacements were averaged to eliminate cubic and higher odd-order anharmonicity, after which the remaining force constants were generated by applying icosahedral symmetry operations. The resulting 180×180 force constant matrix was then diagonalized, and the mode symmetry types were determined by applying group theoretic projection operators to the eigenvectors. Analogous simplifications, based on the high symmetry of C_{60}, have been used in the density functional perturbation theory plane-wave calculations of [2.47]. A different approach is to employ projection operators from the outset, to obtain vibrational symmetry coordinates [2.50]; however, this method involves computing the first-principles forces for many more configurations. Within some of the first-principles methods, it is also possible to completely bypass the computation of the force constant matrix by using the first-principles forces in molecular dynamics and performing a Fourier transform of the resulting time-dependent displacements. This approach was followed by *Adams* et al. [2.50], and also by *Feuston* et al. [2.58] using Car-Parrinello molecular dynamics. Although much more demanding computationally, since the full LDA electronic structure calculation must be carried out at each time step, this method has the advantage that it is not restricted to small displacements, so that anharmonic dynamics can be explored.

While the overall accuracy of first-principles calculations is impressive, it remains between one and two orders of magnitude less than the experimental accuracy. The residual theoretical uncertainty of 1–3% renders it difficult to predict effects which are sensitive to the occurrence of modes with frequency separations of this magnitude. An example is the isotope-induced change of the $A_g(2)$ pentagonal pinch mode Raman peak in molecular C_{60}, which is sensitive to the location of nearby unperturbed frequencies, as detailed in [2.22] and summarized in Sect. 2.2.4.

Once experimental vibrational frequencies became available, several empirical models were introduced. The force constants in these models are obtained from fits to the Raman-active and infrared-active modes. The resulting frequencies, however, are in no better agreement with experiment than those obtained from first-principles methods. *Jishi* et al. [2.59] used an eight-parameter model fit to the ten Raman frequencies of C_{60}. *Feldman* et al. [2.60] used a seven-parameter model fit to the frequencies of the 14 first-order infrared and Raman active modes of C_{60}. The adiabatic bond charge model, which is successful in reproducing the vibrational properties of tetrahedral semiconductors [2.61], has also been applied to fullerenes [2.62, 2.63]. The average discrepancy with experiment obtained from these calculations is about 5%. If one considers the fact that the comparison with experiment includes the modes that are being fit, its is apparent that empirical methods are significantly worse than first-principles approaches. The reasons for this become clear when one examines the spatial range of the interatomic forces in C_{60}. Jishi et al. were the first to note that a model which includes only nearest-neighbor interactions cannot account for the frequencies of the two A_g modes. Hence their model includes up to third-neighbor force constants. However, a careful analysis by *Quong* et al. [2.51] of their first-principles results shows that interatomic force constants extend much further than this – interactions up to seven neighbors must be included to guarantee convergence of the eigenvalues to within $5\,\mathrm{cm}^{-1}$. Similar conclusions regarding the range of the interatomic interactions in C_{60} can be drawn from the work of *Bohnen* et al. [2.64].

Several theoretical studies have appeared on the intermolecular potential that determines the "external," e.g., phonon modes in crystalline C_{60}. While some of these studies concentrate on the prediction of the stable crystalline structures and phase transitions, a few give phonon calculations [2.65, 2.66]. The intermolecular potential is usually modeled as a combination of Morse, van der Waals and Coulomb potentials. First-principles calculations of the intermolecular potential have appeared only recently [2.67], and they have not yet been applied to the phonon modes.

2.2 Vibrational Spectroscopy of C_{60} Molecules

A basic goal of vibrational studies of C_{60} is the identification of its 46 distinct mode frequencies. The discussion of Sect. 2.1.2 showed that for perfectly icosahedral C_{60} only 14 of these frequencies can be determined using first-order ir absorption and Raman scattering. In Sect. 2.1.3 we suggested that information about the 32 silent modes can be obtained by exploiting deviations from icosahedral symmetry, or by performing second-order Raman scattering and infrared absorption experiments. In addition, a number of other techniques, including inelastic neutron scattering, fluorescence spectroscopy and high-resolution electron energy-loss spectroscopy, provide information about silent modes. Unfortunately, none of these methods – considered individually – yields an unambiguous mode identification. In this section we discuss our best attempt at assigning vibrational frequencies on the basis of all the experimental evidence available. We first summarize our conclusions in Sect. 2.2.1, and then focus on the contributions from different techniques in Sects. 2.2.2–8. For completeness, Sect. 2.2.9 briefly discusses some alternative assignments that are found in the literature.

2.2.1 The Assignment of Active and Silent Modes

Table 2.4 in Sect. 2.1.4 gives our recommended values for the 46 different vibrational frequencies in C_{60}, together with two sets of first principles predictions [2.47, 2.53]. Also shown in parentheses are the discrepancies between the calculated and recommended values. The frequencies of the modes that in our opinion have been clearly identified appear in bold type. For the remaining modes the uncertainty in the assignments varies, and future work may lead to somewhat different assignments. Note that the set of well-identified frequencies includes not only the first-order active ir and Raman modes, but also a significant number of silent modes. Whenever possible, the experimental frequencies were selected from low-temperature data. However, since the data originate from different sources, relative errors of the order of $2\,\mathrm{cm}^{-1}$ are unavoidable. In particular, the widespread use of array detectors for Raman spectroscopy leads to discrepancies of the order of $2\,\mathrm{cm}^{-1}$ between different sets of data in the literature, probably as a result of small differences in the detector calibrations.

Several alternative assignments of vibrational frequencies in C_{60} can be found in the literature [2.68, 2.69, 2.70, 2.71, 2.72, 2.73, 2.74, 2.75, 2.76]. *Martin* et al. [2.70] combined their own high-resolution first- and second-order ir data with published Raman measurements [2.68] and neutron scattering results [2.77] to arrive at a complete set of frequencies. Some of the frequencies proposed in [2.70] agree with those in Table 2.4. Others, however, differ from the theoretical predictions given there by as much as 70%. These large discrepancies seem unlikely in view of the remarkable accuracy (better than 5%) of state-of-the-art first-principles methods. In fact, the requirement of consistency with theory can be a powerful tool for the assignment of mode frequencies. This was recognized by *Dong* et al. [2.68] and *Wang* et al. [2.69], who used an empirical model to calculate the mode frequencies. Their frequencies were then adjusted so that the fine structure observed in the Raman and ir spectra of C_{60} could be explained in terms of second-order overtones and combinations. The resulting frequency assignments [2.69] are closer to ours (Table 2.4), but some silent mode frequencies differ by as much as 20%. This discrepancy is understandable if one recalls that empirical vibrational models do not treat active and silent modes on an equal footing. Instead, the known frequencies of the active modes are used as input parameters to generate the interatomic potentials. This makes it difficult to assess the accuracy of the silent mode frequencies predicted by these models. In contrast, for first-principles methods there is no *a priori* reason why the silent mode frequencies should be less accurate than those of the active modes. Since the agreement between theoretical predictions and active mode frequencies turns out to be better than 5%, silent mode frequency assignments which differ from the first-principles predictions by much more than 5% are questionable.

Assignments of silent mode frequencies which are consistent with the density functional perturbation theory calculations of *Giannozzi* and *Baroni* [2.47] were published by *Schettino* et al. [2.72] and by *Menéndez* and *Guha* [2.71]. The latter authors observed that the error in the predicted active mode frequencies has a relatively smooth frequency dependence. The calculations underestimate the low frequencies, while overestimating the high frequencies, as can be seen in Table 2.4. By making the *ad hoc* assumption that the "error curve" for the active modes also applies to the silent modes, *Menéndez* and *Guha* [2.71] generated a set of "adjusted first-principles" frequencies which they used as the starting point for their assignments. The agreement between the assigned frequencies and the adjusted first-principles values is quite remarkable, with a standard deviation of only $2\,\mathrm{cm}^{-1}$. Subsequent neutron scattering experiments [2.40, 2.78], have confirmed all of the assignments of [2.71] for frequencies below $700\,\mathrm{cm}^{-1}$. The recommended frequencies for several of the modes above $700\,\mathrm{cm}^{-1}$ in Table 2.4 are somewhat different from those of [2.71]. The revised frequencies given here are in better agreement with the latest inelastic neutron scattering experiments [2.40, 2.78, 2.79], as well as with recent optical emission experiments [2.80].

2.2.2 First-Order Infrared Absorption and Raman Scattering Experiments

Shortly after the discovery of a method to produce C_{60} in macroscopic quantities [2.2], the observation of four strong infrared absorption peaks provided strong evidence for the proposed icosahedral structure of the molecule [2.81]. With regard to Raman scattering, the unequivocal observation of the ten first-order peaks allowed by group theory required additional advances in the extraction of C_{60} from the fullerene soot. This was because some of the first-order Raman peaks of C_{60} are very weak and could be confused with weak lines arising from C_{70} or other contaminants. Only after chromatographically separated fullerenes became available was it possible to obtain unambiguous evidence for the consistency of the Raman spectrum with the icosahedral symmetry of C_{60} [2.82, 2.83]. Figures 2.1 and 2.2 show infrared absorption and Raman scattering spectra from C_{60} films [2.84]. The measured values of the first-order ir and Raman frequencies are given in Table 2.4.

2.2.3 Second-Order Infrared Absorption and Raman Scattering Experiments

Infrared absorption and Raman scattering involving two vibrational quanta play a key role for the mode frequency assignments. This is because, as dis-

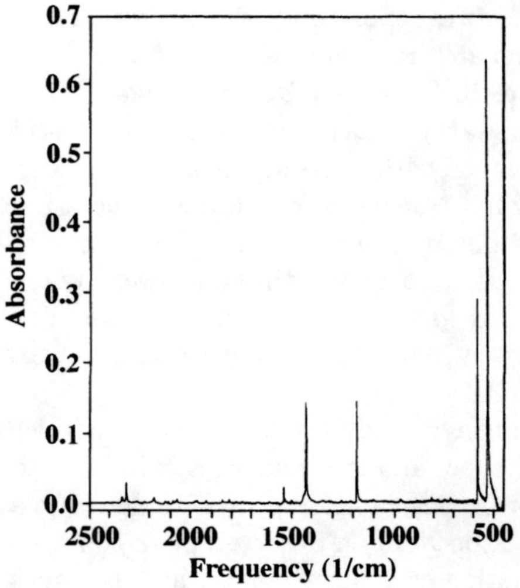

Fig. 2.1. Fourier transform ir spectra of a 1.4 mm thick film of C_{60}. After *Chase* et al. [2.84]

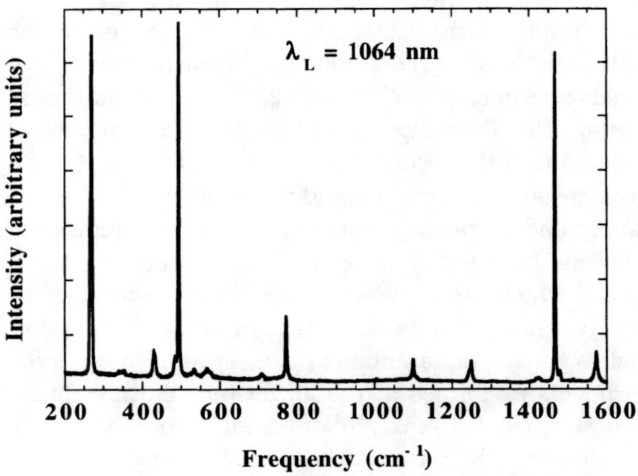

Fig. 2.2. Room temperature Fourier transform Raman spectrum of a C_{60} film. The laser excitation wavelength is 1064 nm. After *Chase* et al. [2.84]

cussed in Sect. 2.1.2, the corresponding selection rules are significantly less restrictive than for first-order absorption or scattering. Table 2.1 listed these selection rules for molecular C_{60}, and it is seen that all of the vibrational symmetries can participate in second order processes. For summation processes, this leads to the possibility of a total of 511 different second-order Raman peak frequencies and 380 different second-order ir frequencies. Notice from Table 2.1 that the inversion symmetry of the molecule imposes the restriction that in Raman summation processes, the two modes must be both even or both odd, whereas in ir they must have opposite parity.

Second-order Raman scattering in C_{60} was reported as early as 1991 [2.85]. A systematic study of all peaks between $250\,\mathrm{cm}^{-1}$ and $3400\,\mathrm{cm}^{-1}$ was given by *Dong* et al. [2.68]. In addition to the 10 first-order Raman peaks of C_{60}, these authors identified 90 additional weaker peaks between $343\,\mathrm{cm}^{-1}$ and $3385\,\mathrm{cm}^{-1}$. Several other groups also studied the fine structure in the C_{60} Raman spectrum [2.73,2.86,2.87,2.88,2.89]. For the second-order ir spectrum of C_{60}, systematic studies were carried out by *Martin* et al. [2.70], *Wang* et al. [2.52], and *Bowmar* et al. [2.73]. The data of Martin et al. were taken at 77 K, whereas the other two sets correspond to room temperature. The agreement between the three sets is good, and that between Martin et al. and Wang et al. is excellent.

Despite the fact that fine structure in ir and Raman spectra provides a powerful tool for the assignment of silent mode frequencies, a failure to explain a few peaks as second-order processes cannot be used to rule out a particular frequency assignment. This is because an observed peak could be either a third-order (or higher) combination or a silent mode which becomes weakly active due to an external perturbation. Both possibilities are known to occur. Table 2.4 indicates that the highest energy vibration in C_{60} is the $H_g(8)$ mode at $1575\,\mathrm{cm}^{-1}$, implying that the highest possible frequency for a second-order peak is $3150\,\mathrm{cm}^{-1}$. However, Raman and ir peaks well above this frequency are observed experimentally [2.52,2.68,2.70], so that they must be due to higher-order processes. Since there is no reason for such processes to be limited to frequencies above the second-order cutoff at $3150\,\mathrm{cm}^{-1}$, the possibility of their occurrence has to be kept in mind when analyzing all of the fine structure in the Raman and ir spectra. Similarly, the lowest vibrational frequency in the C_{60} molecule is the $H_g(1)$ mode at $272\,\mathrm{cm}^{-1}$, setting a lower limit of $544\,\mathrm{cm}^{-1}$ for peaks which can be assigned to second-order spectra (provided that difference processes can be neglected, as is the case at low temperatures). Nevertheless, extra Raman and ir peaks have been observed below $544\,\mathrm{cm}^{-1}$ in experiments performed at the lowest temperatures [2.68, 2.70], and they must be ascribed to weakly allowed first-order processes. Possible mechanisms are discussed in Sects. 2.1.2 and 2.2.4. Again, there is no *a priori* reason why the observation of "forbidden" fundamentals should be limited to modes below $544\,\mathrm{cm}^{-1}$. Therefore the possibility of such processes above this frequency should also be kept in mind.

As a result of the different alternative explanations and the observation of fewer peaks than allowed by group theory, the interpretation of second-order spectra is hardly unique. From the 90 additional Raman peaks identified by *Dong* et al. [2.68], 86 have frequencies equal to or less than $3150\,\mathrm{cm}^{-1}$. A total of 83 such peaks can be explained as allowed second-order combinations or as forbidden first-order processes (i.e., first-order Raman scattering by silent modes), using the assignments of Table 2.4. The standard deviation between the frequencies so computed and the experimental frequencies is only $0.3\,\mathrm{cm}^{-1}$. Three peaks observed by *Dong* et al. [2.68], at $2463\,\mathrm{cm}^{-1}$,

$2782\,cm^{-1}$, and $3118\,cm^{-1}$, cannot be assigned to second-order processes on the basis of the mode frequencies in Table 2.4. These observed peaks are at least one order of magnitude weaker than the strongest second-order lines – which typically involve combinations of H_g modes – and they have not been reported by other groups. All three of these peaks can be interpreted as third-order combinations involving H_g modes, so that their observation cannot be used to rule out the assignments of Table 2.4.

We have also verified that a total of 97 extra infrared absorption peaks observed by *Martin* et al. [2.70] and *Wang* et al. [2.52] can be explained either as second-order combinations or as forbidden fundamentals, using the mode frequency assignments in Table 2.4. The standard deviation of $0.3\,cm^{-1}$ is the same as for the Raman fine structure.

2.2.4 Isotopic and Crystalline Perturbations: Spectroscopic Evidence

The spectroscopic activation of silent modes by isotopic and crystal field perturbations provides an important tool for the study of the vibrational properties of C_{60}. The discussion in Sect. 2.1.3 shows that the two perturbations behave quite differently: isotopes remove the inversion symmetry, thus allowing modes to be both Raman and infrared active in first order. On the other hand, the inversion operator is an element of T_h, so that as long as the T_h symmetry of $\vec{k} = 0$ is preserved, only even modes will be observable with first-order Raman spectroscopy and only odd modes will be first-order ir active. High resolution ir and Raman experiments show peaks at most of the silent mode fundamental frequencies in Table 2.4. On the basis of this table, however, allowed second-order combinations can be predicted to occur near many of these frequencies. Therefore, only those extra peaks observed at the low end of the spectrum, where no second-order combinations are possible, can be ascribed with certainty to silent modes. In experiments on solid C_{60} films at low temperatures, *Dong* et al. [2.68] reported Raman peaks at all of the mode frequencies below $500\,cm^{-1}$ in Table 2.4. Peaks at the same frequencies (with the exception of the frequencies of the $H_g(1)$ and $A_g(1)$ modes) were also observed in ir experiments on similar films by *Martin* et al. [2.70]. This indicates that Raman and ir experiments do not discriminate strongly between even-parity and odd-parity silent modes, suggesting that the isotopic perturbation might be the most important symmetry breaking mechanism. However, systematic investigations of the isotope effect have failed to show any strong correlation between isotopic impurity concentrations and the Raman or ir intensities of silent modes [2.34, 2.89].

The lack of correlation between the intensity of Raman and ir peaks assigned to silent modes and the concentration of ^{13}C isotopes is consistent with our calculations of these intensities using the methods described in Sects. 2.3 and 2.4 [2.90]. These calculations show that isotopes play a very minor role in activating silent modes, so that their observation must primarily be due to crystal field effects. Disorder that reduces the $\vec{k} = 0$ symmetry from T_h, as discussed in Sect. 2.1.3, must play an important role, since it provides the only possible explanation for the removal of the inversion symmetry revealed by the experiments. Nevertheless, one might expect those silent modes which become allowed under T_h to give stronger signals than those which are allowed only due to disorder in the crystalline phases. While a systematic investigation is lacking, some experimental evidence supports this idea. For example, Raman peaks which according to Table 2.4 correspond to G_g modes systematically appear in Raman spectra reported by several different authors [2.68, 2.84, 2.86, 2.87, 2.89]. As shown in Table 2.3 for the fcc ($Fm\bar{3}$) structure, G_g modes are unique in that they belong to the only icosahedral irreducible representation (other than A_g) which contains the identity representation in the decomposition products under T_h symmetry.

Our discussion so far has concentrated on the activation of silent modes by perturbations which lower the icosahedral symmetry. Information about these perturbations can also be obtained by studying their effect on spectroscopically active modes. In fact, it became clear from the earliest Raman studies of C_{60} that the depolarization ratio for H_g modes falls systematically below the value 0.75 predicted on the basis of icosahedral symmetry. The observed depolarization ratio for the two A_g modes is much closer to the predicted value of zero. No detailed explanation has been provided. *Snoke* and *Cardona* [2.91] argued that whereas a general perturbation ϵ can change the depolarization ratio for H_g modes by an amount proportional to ϵ, the change for A_g modes must necessarily be at least quadratic in ϵ, since the unperturbed predicted value of zero is the lowest possible for a depolarization ratio.

The discontinuous temperature dependence of the Raman and ir mode frequencies near the ordering phase transition, together with the splitting of these modes at low temperatures, are strong manifestations of crystal field effects on the intramolecular vibrations of C_{60} [2.92, 2.93, 2.94, 2.95, 2.96]. In most cases, however, these splittings are not well understood. Part of the difficulties stem from the comparable size of the splittings induced by isotopic and crystal-field perturbations, which dictate the need for isotopically pure samples in order to investigate crystal field effects. Perhaps more serious is the possibility of accidental degeneracies between active and silent modes. For closely spaced modes, any small perturbation may induce substantial mixing, so that the analysis of crystal field splittings of active modes requires knowledge of the frequency of all nearby modes. A good example of these complications is provided by the crystal field splittings of the T_{1u} modes. High resolution infrared reflectance experiment by *Homes* et al. [2.94] revealed a complicated splitting pattern for all T_{1u} modes except $T_{1u}(2)$. This pattern was found to be strikingly different in isotopically pure samples [2.95]. The analysis of the splittings was based on the assumption that no silent modes have a frequency close to any of the T_{1u} modes. Our assignments in Table

2.4, however, indicate the presence of silent modes very close in frequency to $T_{1u}(1)$, $T_{1u}(2)$ and $T_{1u}(4)$, a fact that may require a new interpretation of some of the observed splittings.

A more direct way of studying crystal field effects is to perform spectroscopic studies of those external lattice modes which are due exclusively to the weak intermolecular potential. The external modes of the C_{60} crystal have frequencies below 60 cm^{-1}, which makes them difficult to observe. However, an accurate mapping of the phonon dispersion relations has been obtained with inelastic neutron scattering [2.23, 2.97]. The zone-center modes are expected to be optically active and have the convenient feature that those which correspond to translational vibrations are ir active and those which correspond to librational motions are Raman active [2.11]. Both types of modes have been observed [2.98, 2.99, 2.100], and their frequencies are consistent with the neutron scattering experiments.

2.2.5 Quantitative Assessment of Isotope Effects

In previous work [2.21,2.22,2.26], we showed that detailed experimental/theoretical studies of isotopic perturbations on the Raman active modes of C_{60} provide a sensitive probe of the nearby silent modes. Here we summarize aspects of that work, incorporating our present mode assignment of Table 2.4.

Specifically, we analyzed the effects of ^{13}C isotopes on high-resolution Raman spectra in the vicinity of the intense first-order $A_g(2)$ peak that occurs at 1471 cm^{-1} in the isotopically pure icosahedral molecule.[4] This is the familiar "pentagonal pinch" mode, in which the atoms vibrate almost purely tangentially, towards or away from the centers of the 12 pentagons. With isotopes present, the molecule's adiabatic electronic states are unchanged (we are neglecting small zero-point effects), leaving the force-constant matrix, the ir effective charges, and the Raman polarizabilities unaffected. However, the lowered symmetry introduced by mass perturbations relaxes the spectroscopic selection rules and allows the unperturbed modes to couple. Since the perturbation introduced by each ^{13}C isotope in C_{60} is only of order 1/720, the effects on the dynamics are indeed weak, although readily observable in Raman or ir.

The 1.1% natural abundance of ^{13}C means that a C_{60} molecule has a 49% probability of containing one or more of these isotopes. Figure 2.3 shows our high-resolution spectrum for a sample having the natural isotopic abundance. The C_{60} was dissolved in CS_2, and the spectrum was measured at $T = 30$ K. The solid curve is a 3-Lorentzian fit to the experimental points. As detailed

[4] Low-resolution Raman experiments which do not resolve any isotopic structure yield a frequency of 1470 cm^{-1} for the $A_g(2)$ peak, as listed in Table 2.4. The value 1471 cm^{-1} used here corresponds to the $A_g(2)$ mode in $^{12}C_{60}$, which is seen as the high-energy peak in Fig. 2.3.

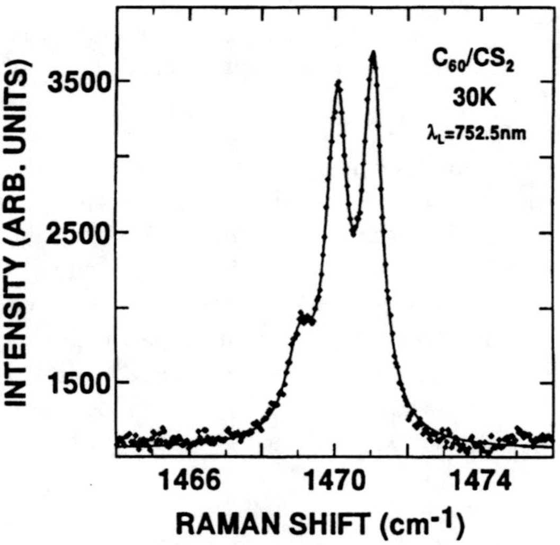

Fig. 2.3. Unpolarized Raman spectrum in the frequency region of the pentagonal pinch mode, for a frozen sample of non-isotopically enriched C_{60} in CS_2 at 30 K. The points represent the experimental data, while the solid curve is a 3-Lorentzian fit. The highest-frequency peak is assigned to the totally-symmetric pentagonal-pinch $A_g(2)$ mode in isotopically pure $^{12}C_{60}$. The other two peaks are assigned to the perturbed pentagonal-pinch mode in molecules having one and two ^{13}C isotopes, respectively. After *Guha* et al. [2.22]

in [2.21,2.22,2.26], we attribute the three clearly-resolved peaks to scattering from molecules containing 0, 1, or 2 ^{13}C atoms, respectively. The measured separation between the first and second peaks is $0.98 \pm 0.01\,\mathrm{cm}^{-1}$, while for the second and third peaks, it is $1.02 \pm 0.02\,\mathrm{cm}^{-1}$. As shown below, these splittings are predicted essentially exactly by first-order nondegenerate perturbation theory.

The bottom panel of Fig. 2.4 shows the analogous spectrum, but for an isotopically enriched sample. The top panel gives the measured isotopic mass distribution for this sample. It is apparent that the two lineshapes show a remarkable correlation. Again, a very simple perturbation-theoretic argument will be seen to explain this result, while deeper analysis will correlate this success with information on the nearby unperturbed frequency positions.

Given the smallness of the isotopic mass perturbations, it is natural to try perturbation theory, the simplest form of which is nondegenerate perturbation theory. With isotopes present, the mass matrix in the mode eigenvalue problem (2.5) becomes $\boldsymbol{M}_0 + \Delta\boldsymbol{M}$, where $\boldsymbol{M}_0$ corresponds to the unperturbed molecule. As noted above, the force constant matrix remains unperturbed. The derivation of the perturbation series is straightforward, although the problem is slightly different from the usual case based on $(H_0 + V)\psi = E\psi$, because here the perturbation $\Delta\boldsymbol{M}$ is a multiplier of the eigenvalue and also appears in the eigenvector normalization (2.7). A derivation of the perturbation series for the frequencies which avoids the normalization condition is given in [2.26]. Here we list the relevant results.

The perturbed frequency for a nondegenerate mode f is given by the series

$$\frac{\Delta\omega_f^2}{\omega_{0f}^2} = \sum_{n=1,2,\ldots} \left(\frac{\Delta\omega_f^2}{\omega_{0f}^2}\right)_n , \tag{2.29}$$

where $\Delta\omega_f^2 \equiv \omega_f^2 - \omega_{0f}^2$ is the shift of the squared frequency of the mode. The first three terms in the right-hand side of this equation are

$$\left(\frac{\Delta\omega_f^2}{\omega_{0f}^2}\right)_1 = -\tilde{\chi}_0(f)\Delta \boldsymbol{M}\chi_0(f), \tag{2.30}$$

$$\left(\frac{\Delta\omega_f^2}{\omega_{0f}^2}\right)_2 = (\Delta\mathcal{M}_{ff})^2 + \sum_{f'\neq f} \frac{(\Delta\mathcal{M}_{ff'})^2}{1-(\frac{\omega_{0f'}}{\omega_{0f}})^2} , \tag{2.31}$$

and

$$\begin{aligned}\left(\frac{\Delta\omega_f^2}{\omega_{0f}^2}\right)_3 = & -(\Delta\mathcal{M}_{ff})^3 - 3\Delta\mathcal{M}_{ff}\sum_{f'\neq f}\frac{(\Delta\mathcal{M}_{ff'})^2}{1-(\frac{\omega_{0f'}}{\omega_{0f}})^2} \\ & +\Delta\mathcal{M}_{ff}\sum_{f'\neq f}\frac{(\Delta\mathcal{M}_{ff'})^2}{\left[1-(\frac{\omega_{0f'}}{\omega_{0f}})^2\right]^2} \\ & -\sum_{f',f''\neq f}\frac{\Delta\mathcal{M}_{ff'}\Delta\mathcal{M}_{f'f''}\Delta\mathcal{M}_{f''f}}{\left[1-(\frac{\omega_{0f'}}{\omega_{0f}})^2\right]\left[1-(\frac{\omega_{0f''}}{\omega_{0f}})^2\right]} ,\end{aligned} \tag{2.32}$$

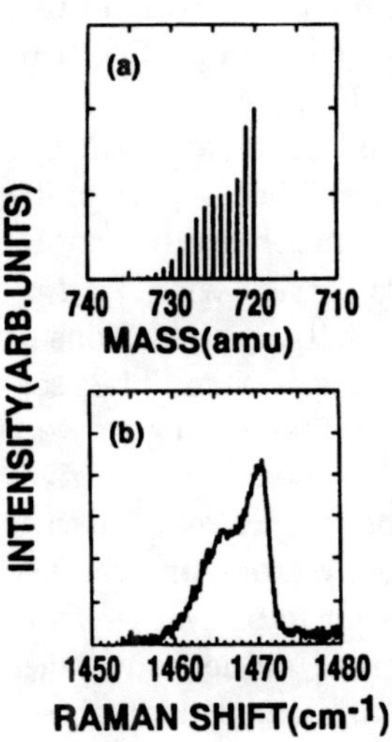

Fig. 2.4. (**a**) Mass spectrum of a ^{13}C-enriched C_{60} sample. (**b**) Unpolarized Raman spectrum of the same enriched sample in CS_2 at 30 K. After *Guha* et al. [2.21]

where $\Delta \mathcal{M}_{ff'} \equiv \tilde{\chi}_0(f) \Delta \boldsymbol{M} \chi_0(f')$ is the isotope perturbation matrix $\Delta \boldsymbol{M}$, expressed in the basis of unperturbed mode eigenvectors. While just the first two terms are needed in the following, the third-order result is included for reference.

For A_g modes in C_{60} molecules, the first-order shift (2.30) reduces to a very simple result. Because every site in the unperturbed icosahedral molecule is symmetry equivalent, the magnitudes of the atomic displacements for the unperturbed totally symmetric A_g modes are site-independent, and for the case of n isotopic substitutions $^{12}C \to {}^{13}C$, (2.30) gives

$$\left(\frac{\Delta \omega_{A_g}^2}{\omega_{0A_g}^2} \right)_1 = -\frac{n}{720} \, . \tag{2.33}$$

This nondegenerate perturbation theory result is independent of the location of the isotopes and is valid provided the unperturbed A_g mode is sufficiently far from the nearby unperturbed modes to which it couples via the perturbation. For the $n = 1$ case of a single isotope, the result is $\Delta \omega_{A_g} = \omega_{0A_g}(\sqrt{719/720} - 1)$, and for the unperturbed $A_g(2)$ mode at $1471\,\mathrm{cm}^{-1}$, the predicted shift is $-1.02\,\mathrm{cm}^{-1}$, in excellent agreement with the measured shift of $-0.98\,\mathrm{cm}^{-1}$.

The above first-order perturbation theory result is also consistent with the strong correlation between the Raman and mass spectrum lineshapes seen in Fig. 2.4. From (2.24), the strength of the Raman scattering from mode f is determined by the mode polarizability derivatives $P_{\alpha\beta,f} = \sum_{\ell\gamma} P_{\alpha\beta}(\ell\gamma)\chi(\ell\gamma|f)$. The real-space electronic polarizability derivatives $\{P_{\alpha\beta}(\ell\gamma)\}$ are not changed under isotopic substitution, as noted above, and when first-order perturbation theory is valid, the mode eigenvectors $\{\chi(\ell\gamma|f)\}$ remain unperturbed as well. The Raman intensity for the perturbed A_g mode will then be the same as for the unperturbed molecule. Hence as long as the first-order perturbation theory result (2.33) remains valid, the Raman spectrum for a mixture of C_{60} molecules having varying amounts of ^{13}C should simply be proportional to the mass spectrum of the sample, exactly as is seen in Fig. 2.4.

As the number of isotopes is increased, (2.33) should eventually fail, but the good agreement just noted implies that the unperturbed $A_g(2)$ mode frequency is relatively isolated from the frequencies of nearby unperturbed modes which can couple appreciably through the mass perturbation. To further explore this question, we also performed nonperturbative calculations of the perturbed normal modes and the associated Raman spectrum. The real-space Raman polarizability derivatives needed for the Raman strengths were obtained from a phenomenological bond-polarizability model which gives an excellent account of the unperturbed C_{60} off-resonance Raman intensities [2.22], as we detail in Sect. 2.4.1. For the mode eigenvectors, one can use (2.5) directly, provided the force constant matrix $\boldsymbol{\Phi}$ is available, either from first-principles or phenomenological models. The resulting mode frequencies and eigenvectors can then be computed for any spatial arrangement of iso-

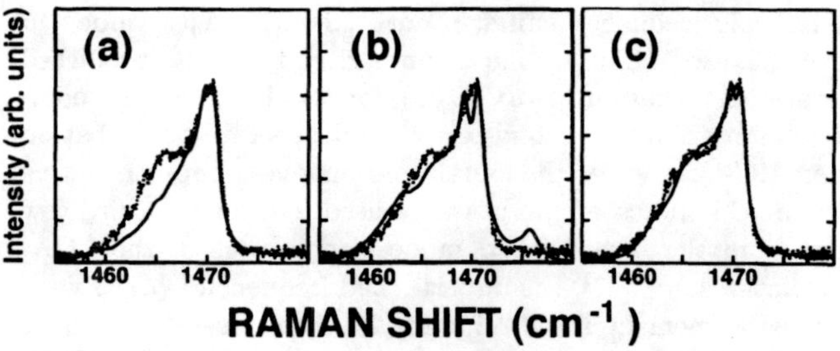

Fig. 2.5. Solid lines: calculated Raman spectra in the frequency region of $A_g(2)$ for the ^{13}C-enriched C_{60} of Fig. 2.4. Up to 10 isotopes were included, and the spectra were weighted according to their relative abundances in the mass spectrum. These spectra were calculated nonperturbatively, using the first-principles eigenmodes of [2.53] for panel (**a**), the first-principles eigenmodes of [2.47] for panel (**b**), and our present frequency assignment of column 2 in Table 2.4 for panel (**c**), by implementing (2.34) in the manner described in the text. The dashed line is the experimental Raman spectrum of Fig. 2.4b

topes, and the corresponding Raman spectra predicted. Our results for such calculations, based on the first-principles LDA-based local-orbitals technique discussed in Sect. 2.1.4, together with the isotopic mixture shown in the top panel of Fig. 2.4, are given in panel (a) of Fig. 2.5. The distribution encompasses 0 to 10 ^{13}C isotopes per molecule, and as detailed in [2.22], the predicted spectra in Fig. 2.5 include exactly all possible isotopic arrangements for the cases of 0, 1, and 2 ^{13}C isotopes, whereas the cases of 3–10 isotopes are included within an accurate approximation. The agreement between the predicted and observed spectrum is seen to be quite good, although not as good as that found with the simple first-order perturbation theory argument given above, which bypasses detailed model calculations altogether.

Further insight into this problem is obtained by using the corresponding unperturbed mode frequencies and eigenvectors in (2.31) to evaluate the second-order nondegenerate perturbation theory shifts. The result for the $n = 1$ case of a single isotope is listed in the third column of Table 2.5, together with the exact nonperturbative shift predicted by our first-principles model, given in the fourth column. As anticipated from the simple first-order nondegenerate perturbation theory results, the second-order shift is significantly smaller than the first-order shift. Nevertheless, it is interesting that the first-order nondegenerate perturbation theory result was in fact essentially perfect, in which case the perturbation theory contributions beyond first order should vanish.

In view of the high accuracy of *Giannozzi* and *Baroni*'s first-principles density functional perturbation theory predictions for the frequencies of icosahedral C_{60} [2.47], displayed here in Table 2.4, we have repeated our isotopically perturbed Raman spectrum calculations, using their model. The resulting predicted spectrum for the enriched sample of Fig. 2.4 is shown in

Table 2.5. Calculated isotopic frequency shifts, in cm^{-1}, for the $A_g(2)$ mode in molecular C_{60} having one ^{13}C atom. The second column gives the shifts predicted by first-order nondegenerate perturbation theory (2.33), for $n = 1$. They are model-independent. The third column gives the second-order nondegenerate perturbation theory shifts obtained from (2.31), whereas the fourth column gives nonperturbative shifts. For the first two rows, the indicated models were used, but for the third row $\{\Delta\mathcal{M}_{ff'}\}$ was computed from the unperturbed mode eigenvectors of the LDA-based first-principles calculations of [2.47], the unperturbed frequencies $\{\omega_{0f}\}$ were taken from our frequency assignment in Table 2.4, and (2.34) was used to compute the nonperturbative shifts, as explained in the text. The perturbation theory shifts of the second and third columns were obtained using the experimental Raman frequency $\omega_0 = 1471\,\mathrm{cm}^{-1}$ in the left-hand sides of (2.30) and (2.31)

Frequency set	1st-order shift (model independent)	2nd-order shift	Nonperturbative shift
Adams et al.[a]	−1.02	0.21	−0.86
Giannozzi and *Baroni* [b]	−1.02	−0.33	−1.38
Present assignment[c]	−1.02	0.015	−0.97
Experimental shift: −0.98			

[a] Table 2.4, column 4 (from [2.53])
[b] Table 2.4, column 3 (from [2.47])
[c] Table 2.4, column 2

panel (b) of Fig. 2.5, and it is seen to be in very good agreement with the experimental spectrum. The second-order nondegenerate perturbation theory shift computed using Giannozzi and Baroni's model is given in the second row of Table 2.5, along with the predicted exact shift. Like our results in the first row, these results are qualitatively consistent with the earlier success of simple first-order nondegenerate perturbation theory, but are quantitatively worse than that simple model-independent prediction.

At this point it is useful to recast the fundamental mode eigenvalue problem (2.5) into an equivalent form which more transparently brings out the importance of the lineup of unperturbed frequencies. This form is simply obtained by re-expressing (2.5) in the complete basis of unperturbed orthonormal mode eigenvectors $\{\chi_0(f)\}$. The result is

$$\sum_{f'}[\omega_{0f}^2\delta_{ff'} - \omega_f^2(\delta_{ff'} + \Delta\mathcal{M}_{ff'})]c_{ff'} = 0, \tag{2.34}$$

where the matrix elements $\Delta\mathcal{M}_{ff'}$ are defined just below (2.32), and the coefficients $\{c_{ff'}\}$ are the expansion coefficients of the mode eigenvectors in the basis of unperturbed eigenvectors:

$$\chi(f) = \sum_{f'=1}^{3N} c_{ff'}\chi_0(f'). \tag{2.35}$$

The differences in the predicted isotopically perturbed Raman spectra of panels (a) and (b) of Fig. 2.5 derive from differences in the mode frequencies and eigenvectors for the two first-principles methods used. We therefore compared the frequencies and eigenvectors for these two models mode by mode and found that while the frequencies differ by as much as 7%, their eigenvectors are practically identical. This distinction between the behavior of the frequencies and eigenvectors in the two models is even more striking when one recalls that the mode eigenvalues are actually the *squares* of the frequencies. Apparently, the very high symmetry of C_{60} imposes rather severe constraints on the forms of the mode displacement patterns. For the isotope problem, this means that the matrix elements $\Delta\mathcal{M}_{ff'}$ appearing in (2.34) are highly similar in the two models used for panels (a) and (b) of Fig. 2.5. Hence the dominant contribution to the differences in the two predicted isotopically perturbed spectra derives from the models' different frequency lineups in pure C_{60}.

Moreover, we can use (2.34) to study easily the effect of our frequency assignment in Table 2.4 on the isotopically perturbed Raman scattering, even though we have no force constant model underlying these frequencies. This is accomplished by using column 2 of Table 2.4 for the ω_{0f}'s and simply evaluating the matrix elements $\Delta\mathcal{M}_{ff'}$ using the eigenvectors obtained from either of the above two models. The resulting isotopically perturbed Raman spectrum is given in panel (c) of Fig. 2.5, and it is in excellent agreement with experiment. In addition, we can use the same $\Delta\mathcal{M}_{ff'}$'s to compute the second-order nondegenerate perturbation theory shift from (2.31), yielding the result in the third row of Table 2.5. Now the second-order shift is indeed negligible, as anticipated.

In [2.22], we performed these approximate, but nonperturbative, calculations of the isotopically perturbed $A_g(2)$ Raman spectrum, for several other published models of the normal modes of pure C_{60}, for which the frequencies, but not the eigenvectors, were available. As shown there, models with unperturbed silent modes having frequencies very close to $A_g(2)$ yield predicted isotopically perturbed spectra which poorly represent the experimental spectrum of Fig. 2.4b. Moreover, the predicted shifts are significantly off, and the second-order nondegenerate perturbation contributions turn out to be larger than the first-order corrections. Reference [2.22] should be consulted for details. Our results demonstrate that this sort of isotopically perturbed Raman scattering study can provide a sensitive probe for quantitative assessments and refinements of theoretical models.

2.2.6 Inelastic Neutron Scattering

Inelastic neutron scattering (INS) is the most comprehensive technique for the study of vibrations in condensed matter: the $\vec{k}$ conservation and other selection rules are much less restrictive than for optical spectroscopies. In the case of C_{60}, all modes, including those silent for first-order Raman and

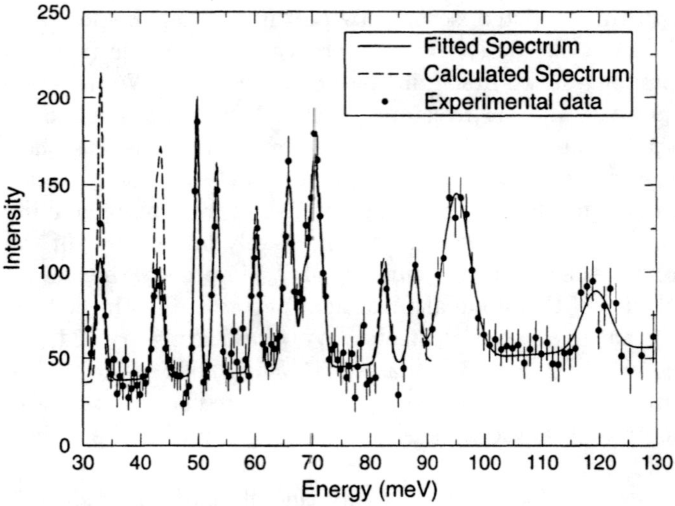

Fig. 2.6. Measured inelastic neutron scattering spectrum of C_{60}. The solid line shows a fit to a series of Gaussians plus a sloping background. A calculation using frequencies that are virtually identical with those of Table 2.4 is shown as a solid line. Note that the calculated spectrum ends at 90 meV (730 cm^{-1}). The excellent agreement between the calculation and the Gaussian fit can be extended to 130 meV (1050 cm^{-1}), using the frequencies in Table 2.4. From [2.78].

infrared spectroscopy, can be detected with this technique. The scattering intensity is proportional to the mode degeneracy, a fact that provides a helpful tool for the identification of the vibrations. The primary disadvantage of INS is its poor spectral resolution. This makes it difficult to individually resolve all of the modes, especially when several frequencies are clustered within a narrow range, as is the case for many C_{60} vibrations.

The initial inelastic neutron scattering studies of intramolecular C_{60} modes provided an overall picture of the density of vibrational states and were reasonably consistent with the existent Raman and infrared data, but the resolution was not good enough to identify other modes, particularly in the high energy range [2.101]. Subsequent work with improved resolution provided the first experimental evidence for several of the assignments in Table 2.4 [2.77, 2.102]. *Prassides* et al. [2.102], for example, identified the $T_{3u}(1)$, $G_u(1)$ and $H_u(1)$ modes. Later work has given a detailed account of all modes below 700 cm^{-1} [2.40, 2.78]. Figure 2.6 shows results by *Copley* et al. [2.78]. The solid line represents a fit to the data with a set of Gaussian functions, whereas the dashed line corresponds to a calculation using mode frequencies that are essentially the same as those in column 2 of Table 2.4 and in the earlier assignment by *Menéndez* and *Guha* [2.71]. The agreement between the fit and the calculation is excellent, except for the intensity at the lowest energies, where the discrepancy has been attributed to coherency effects [2.78].

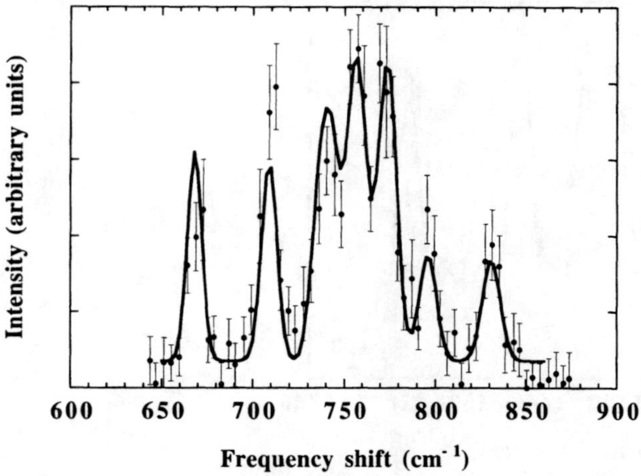

Fig. 2.7. Experimental inelastic neutron scattering spectrum of C_{60}, showing medium-energy modes (circles) [2.79]. The solid line is a calculated spectrum consisting of a sum of Gaussians centered at the mode frequencies of Table 2.4 and weighted according to the modes' degeneracies. The width of each Gaussian is taken to be proportional to the mode's energy and adjusted to the experimental data. The best agreement is obtained using $\Delta E/E = 0.014$. A constant background has been added, and the energies were rescaled as discussed in the text

The agreement between the frequencies of Table 2.4 and the results of INS extends beyond $700\,\mathrm{cm}^{-1}$. Figures 2.7 and 2.8 compare a theoretical prediction based on Table 2.4 with recent INS data obtained by *Coulombeau* et al. [2.79]. For this experiment, the authors used the MARI spectrometer at the ISIS pulsed neutron source of the Rutherford Appleton Laboratory, UK. This spectrometer features an array of detectors distributed over a wide angular range, and consequently the momentum transfer $\hbar\vec{Q}$ is not well defined. Since this makes it difficult to include Debye–Waller and coherency effects in the theoretical simulations, we computed each of the theoretical spectra in Figs. 2.7 and 2.8 as a simple sum of Gaussians with weights proportional to the mode degeneracies and a single width parameter $\Delta E/E$ adjusted to best fit the experimental data. The agreement is excellent, and it is made nearly perfect by rescaling the experimental frequencies by a factor of 0.99. This is shown in Figs. 2.7 and 2.8. The necessity for the rescaling factor is suggested by the fact that some of the INS peaks which originate from spectroscopically active vibrations appear at an energy 1% higher than the value known from ir or Raman spectroscopy. It is quite apparent that the INS data provide strong support for the mode assignments in Table 2.4.

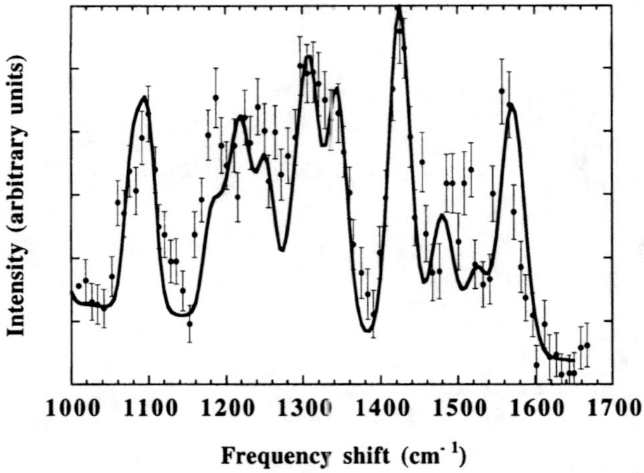

Fig. 2.8. Experimental inelastic neutron scattering spectrum of C_{60}, showing high-energy modes (circles) [2.79]. The solid line is a calculated spectrum consisting of a sum of Gaussians centered at the mode frequencies of Table 2.4 and weighted according to the modes' degeneracies. The width of each Gaussian is taken to be proportional to the mode's energy and adjusted to the experimental data. The best agreement is obtained using $\Delta E/E = 0.022$. A linear background has been added, and the energies were rescaled as discussed in the text

2.2.7 Optical Spectroscopy

Optical absorption spectroscopy, fluorescence and phosphorescence spectra can also be powerful tools for the identification of vibrational modes in molecules. Of particular relevance for our analysis is the work of Sassara et al., who investigated – with exquisite detail – the vibrational fine structure in the fluorescence [2.80] and phosphorescence [2.103] spectra of C_{60} molecules molecules isolated in neon and argon matrices. This work leads to the unambiguous identification of several vibrational modes of T_{3u}, H_u, and G_u symmetry. The major improvements in Table 2.4 relative to the earlier frequency assignments of *Menéndez* and *Guha* [2.71] reflect the contribution of the fluorescence experiments.

According to theoretical calculations [2.45, 2.104], the three lowest singlet excited states in C_{60} are nearly degenerate and belong to the g_g, t_{3g}, and t_{1g} irreducible representations of the icosahedral group. Here we are using lower-case letters for the primary irreducible representation labels to distinguish electronic from vibrational symmetries. Fluorescence from these states is therefore dipole-forbidden, and the corresponding spectrum is dominated by vibrational bands associated with Herzberg–Teller and Jahn–Teller modes. Since the calculated energy separation between the lowest excited states is very small, the actual lineup of singlet states is likely to be strongly dependent on environmental perturbations. This is confirmed by the experimental work of *Sassara* et al. [2.80] Figures 2.9 and 2.10 show their fluorescence

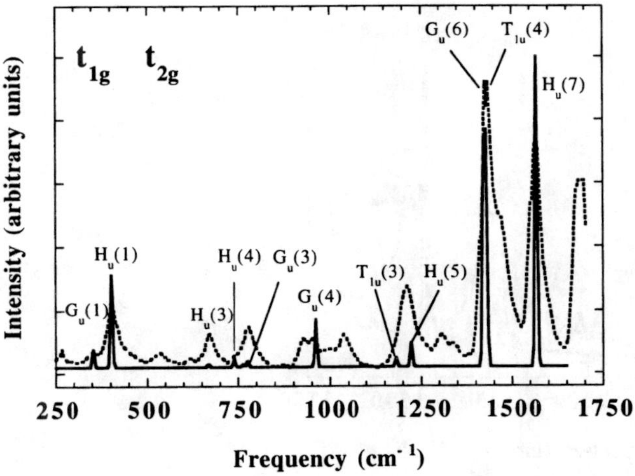

Fig. 2.9. Experimental fluorescence spectrum of C_{60} in a neon matrix (dashed line) [2.80]). The frequency scale represents the shift from the origin of the fluorescence. The solid line represents the intensities of Herzberg–Teller modes calculated by *Negri* et al. [2.107] and plotted according to the mode frequencies in Table 2.4. The individual peaks in the spectrum were simulated as Gaussians with a FWHM of $7\,\mathrm{cm}^{-1}$. An adjustable parameter was introduced to fit the relative intensities of the t_{1g} and t_{3g} manifolds

spectra from C_{60} molecules isolated in neon and argon matrices, respectively. The differences in the spectra reflect the different symmetry of the Herzberg–Teller and Jahn–Teller vibrational modes associated with each of the three low-lying singlet states. Active Herzberg–Teller modes are those that couple the lowest singlet states to higher-energy, dipole-allowed t_{1u} states. It is easy to show that the relevant vibrational mode symmetries are H_u and G_u (t_{3g}); A_u, H_u, and T_{1u} (t_{1g}); and H_u, G_u, and T_{3u} (g_g). For Jahn–Teller modes, the symmetries of the active components are H_g (g_g, t_{3g}, and t_{1g}) and G_g (g_g) [2.105]. These symmetry considerations, combined with calculations of Herzberg–Teller oscillator strengths by *Negri* et al. [2.105], allowed *Sassara* et al. [2.80] to explain the fluorescence spectrum from neon matrices as an admixture of t_{1g} and t_{3g} singlet emission. Similarly, the fluorescence spectrum from argon matrices was assigned to a mixture of t_{1g} and g_g emission. From this analysis it is also possible to determine the frequency of several silent modes of H_u, G_u, and T_{3u} symmetry.

The solid lines in Figs. 2.9 and 2.10 show the predicted fluorescence intensity based on the Herzberg–Teller calculations of *Negri* et al. [2.105, 2.106, 2.107] and our mode frequency assignments from Table 2.4. The only adjustable parameter in the fit is the relative intensity of the emission from the t_{1g} and t_{3g} states (neon matrices) or t_{1g} and g_g states (argon matrices). It is apparent that the fluorescence data provide very strong experimental

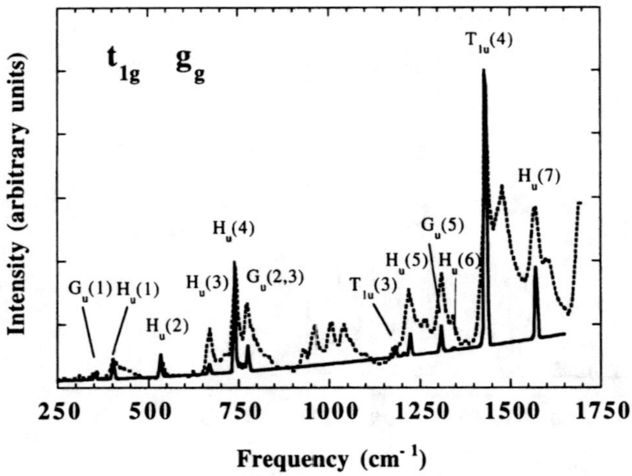

Fig. 2.10. Experimental fluorescence spectrum of C_{60} in an argon matrix (dashed line [2.40]). The frequency scale represents the shift from the origin of the fluorescence. The solid line represents the intensities of Herzberg–Teller modes calculated by *Negri* et al. [2.107] and plotted according to the mode frequencies in Table 2.4. The individual peaks in the spectrum were simulated as Gaussians with a FWHM of 7 cm^{-1}. An adjustable parameter was introduced to fit the relative intensities of the t_{1g} and g_g manifolds. A sloping background was also added to improve the agreement between the calculation and the experimental data

support for many of the assignments in Table 2.4, most notably for modes $H_u(1)$, $H_u(4)$, $G_u(6)$, and $H_u(7)$.

We indicated above that Table 2.4 incorporates the fluorescence results through modification of some of the earlier mode assignments made by *Menéndez* and *Guha* [2.71]. On the other hand, the original frequency assignments in [2.71] and [2.72], together with the first-principles normal mode calculations of *Baroni* and *Giannozzi* [2.47], led to a revised interpretation of some of the details of the fluorescence spectra. This is because *Sassara* et al. based their initial analysis [2.80] on earlier calculations by *Negri* et al. [2.105]. A comparison with experimental data that became available after these calculations were published, as well as with the vibrational frequencies of [2.47], led *Negri* et al. to propose a slightly different assignment of oscillator strengths [2.106]. Additional discrepancies with the frequency assignments of *Menéndez* and *Guha* [2.71] motivated yet another revision [2.107], which has been incorporated into Figs. 2.9 and 2.10. It is gratifying to note that the latest calculations of *Negri* et al. [2.106, 2.107] make it possible to give a unified explanation of the fluorescence and inelastic neutron scattering spectra in terms of the mode frequencies of Table 2.4.

From the point of view of silent mode spectroscopy, the Herzberg–Teller vibronic activity shown in Figs. 2.9 and 2.10 is the most interesting component of the fluorescence. The Jahn–Teller contribution is dominated by

Raman-active H_g modes and therefore offers little additional insight into the vibrational structure of C_{60}. However, this contribution must be computed for a detailed quantitative comparison between experimental and theoretical fluorescence spectra, as recently done by *Sassara* et al. [2.108]. They also included Franck-Condon terms and extended the calculations to take into account two additional many-electron excited states, of t_{3u} and h_g symmetry. The resulting agreement between theory and experiment is very remarkable, with the calculation reproducing the rich fluorescence spectrum, even in its minor details.

2.2.8 High-Resolution Electron Energy Loss

A beam of monoenergetic electrons is scattered by the electric dipole associated with infrared-active vibrations. This leads to the observation of infrared-active modes in high-resolution electron-energy loss (HREELS) measurements. HREELS experiments require an ultra-high vacuum environment and cannot match the resolution of optical measurements, so that there appears to be no obvious advantage in using this technique for the determination of the infrared mode frequencies. However, HREELS experiments are important for the assignment of vibrational frequencies in C_{60} because *all* modes can induce inelastic scattering of electrons via the short-range components of the electron-molecule interaction. This is the so-called impact scattering mechanism [2.109]. *Gensterblum* et al. [2.109,2.110] have performed HREELS experiments on C_{60} films. Depending on the substrate used, the films are crys-

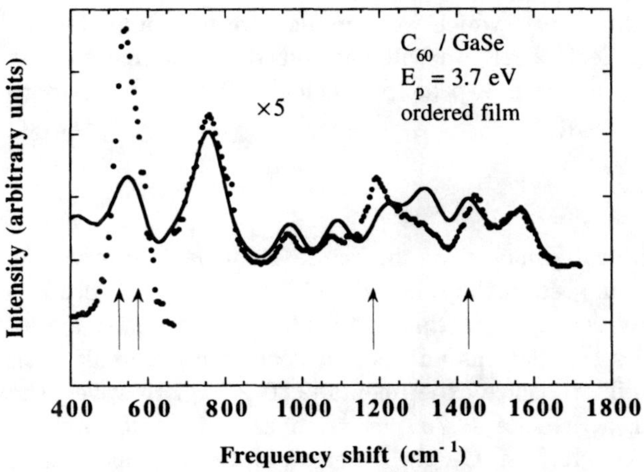

Fig. 2.11. Experimental HREELS for a C_{60} film grown on GaSe(0001) (points). From [2.72]. The solid line represents the vibrational density of states in C_{60}, calculated according to the mode frequencies in Table 2.4. The Gaussian broadening of the density of states was chosen as $70\,cm^{-1}$ (FWHM), which gives an excellent fit of the prominent feature at $758\,cm^{-1}$. The vertical arrows represent the position of the four T_{1u} modes

talline or disordered. This has a significant effect on the relative intensity of the different spectral features. Figure 2.11 shows experimental data for an ordered C_{60} film grown on a GaSe substrate. The spectrum is compared with a broadened density of vibrational states computed from Table 2.4. It is quite apparent that the mode assignment in Table 2.4 is consistent with all of the peaks observed with HREELS. The computed density of states also agrees surprisingly well with the peak intensities for the ordered film. The largest discrepancies correspond to the ir active vibrations, which are indicated by arrows in Fig. 2.11. These discrepancies, however, are not unexpected, since the infrared-active vibrations scatter the electrons via the additional dipole scattering mechanism, which is sensitive to the dipole matrix elements.

According to Table 2.4, 41 modes (including degeneracies) appear in the 660–800 cm^{-1} range. This is consistent with the prominent HREELS peak at 758 cm^{-1}. An equivalent broad structure is apparent in the inelastic neutron scattering spectra. It is noteworthy that within this range there are no infrared-active modes, and the only Raman modes are $H_g(3)$ and $H_g(4)$. This means that the bulk of the structure is due to silent modes. Similarly, other HREELS peaks can be entirely explained in terms of silent modes. For example, a HREELS peak at 347 cm^{-1} (not shown in Fig. 2.11) can be identified with the $T_{3u}(1)$ and $G_u(1)$ modes, and the A_u, $T_{3u}(3)$, and $G_u(4)$ modes in Table 2.4 correspond to the HREELS peak at 960 cm^{-1}.

2.2.9 Alternative Silent Mode Assignments

Table 2.4 shows 17 silent modes for which the frequency we recommend cannot be considered definitive. Several of these modes have been given alternative assignments in the literature, which we compile here for completeness. For reasons discussed in Sect. 2.2.1, we will not consider assignments that deviate from the most accurate first-principles calculations by much more than 5%.

$T_{2u}(2)$

There are two alternative assignments for this mode: an ir peak observed at 709–712 cm^{-1} or an ir peak seen in the range 753–757 cm^{-1} [2.52, 2.70, 2.73]. The first assignment, favored by *Schettino* et al. [2.72], is in better agreement with first-principles calculations and is consistent with a weak peak observed at 714 cm^{-1} in fluorescence experiments [2.80, 2.108]. However, the close proximity of the $H_g(3)$ mode at 709 cm^{-1} makes it difficult to reach definitive spectroscopic conclusions. On the other hand, the ir peak seen in the 753–757 cm^{-1} range cannot be explained as a second-order combination, and it provides, together with our other assignments, the only plausible way to explain the ir peak observed at 2176 cm^{-1} [2.52, 2.70, 2.73], namely as the combination $H_g(7) + T_{2u}(2)$. Turning to the inelastic neutron scattering data, we note that *Copley* et al. [2.78] report a peak at 708 cm^{-1}, which is resolution limited, so that if $T_{2u}(2)$ is part of this structure it would have to be almost degenerate with the $H_g(3)$ mode. The intensity of this INS peak is comparable to that of a nearby INS peak at 668 cm^{-1} [2.78], which corresponds to the 5-fold degenerate $H_u(3)$ mode. Hence, the similar intensity of the 708 cm^{-1} INS peak argues against the existence of both a 5-fold degenerate and a 3-fold degenerate mode at this frequency. Very similar conclusions can be reached from *Coulombeau's* MARI data [2.79]. In addition, the higher resolution of this experiment allows the authors to resolve three peaks in the broad structure around 750 cm^{-1}. The theoretical calculation of the INS spectrum agrees better with this three-peak structure if the T_{2u} peak is placed at 753 cm^{-1}, as shown in Fig. 2.7.

$T_{1g}(2)$

Our assignment of this mode is based on INS data [2.77,2.79,2.102], showing a peak near 840 cm^{-1}. A peak at 813 cm^{-1} has also been reported [2.70]. Both structures have been assigned to $T_{1g}(2)$ [2.71,2.72]. The 840 cm^{-1} assignment is further supported by a Raman peak observed by *Rosenberg* et al. [2.89] at 835 cm^{-1} in ^{13}C enriched samples, although this peak can also be assigned to a second-order combination [2.89]. As can be seen in Fig. 2.7, the agreement of the MARI INS data of *Coulombeau* et al. [2.79] with the frequencies in Table 2.4 is excellent. On the other hand, *Copley* et al. [2.78] fail to observe any evidence for INS peaks in the 800–900 cm^{-1} range. Moreover, they claim that the strong structure between 725 and 800 cm^{-1} should be assigned to four 3-fold degenerate modes, three 4-fold modes, and two 5-fold modes. This agrees exactly with our assignments if we place the $T_{1g}(2)$ mode below 800 cm^{-1}. We noticed that a frequency of 790 cm^{-1} provides a good fit of the Copley et al. neutron scattering data. However, the latest MARI INS data of *Coulombeau* et al. [2.79] probably should be preferred in this range, given the better resolution and improved signal-to-noise ratio. This is the reason for our recommended frequency for $T_{1g}(2)$.

$T_{2u}(3)$

The two first-principles calculations in Table 2.4 predict three modes in the range 900–1000 cm^{-1}. These are the $G_u(4)$, $T_{2u}(3)$, and A_u modes, and this prediction is in good agreement with INS data. *Copley* et al. [2.78] find an INS peak at 970 cm^{-1} which has a total integrated intensity consistent with these three modes. The frequency of the $G_u(4)$ mode is 961 cm^{-1} according to the fluorescence experiments [2.80,2.108]. We obtain an excellent fit of the INS data of [2.78] by further assuming that the frequencies of the $T_{2u}(3)$ and A_u modes are 973 cm^{-1} and 984 cm^{-1}, respectively. Unfortunately, the high-resolution MARI INS data of *Coulombeau* et al. [2.79] do not include this frequency range. Our assignment is also consistent with an observed ir peak at 973 cm^{-1} [2.70], although this peak can also be explained as a second-order combination. An alternative assignment for $T_{2u}(3)$, proposed by *Schettino* et al. [2.72], is the ir peak observed at $1040\,\mathrm{cm}^{-1}$ [2.52, 2.72, 2.73], although this peak can also be explained in terms of the second-order combination $H_g(1)+G_u(2)$. No INS structure is seen at this frequency.

$G_g(4)$

We assigned $G_g(4)$ to a Raman peak observed by several authors at $1079\,\mathrm{cm}^{-1}$ [2.68, 2.86, 2.87, 2.89]. An alternative assignment is a Raman peak seen at 1138–$1142\,\mathrm{cm}^{-1}$ [2.68, 2.73, 2.86]. *Gallagher* and coworkers have observed a strong resonance of this peak and have analyzed its excitation profile on the basis of the $G_g(4)$ assignment [2.111, 2.112]. However, as pointed out by these authors and others, the $1140\,\mathrm{cm}^{-1}$ peak can also be explained in terms of the second-order combination $H_g(2)+H_g(3) = 1142\,\mathrm{cm}^{-1}$. Our assignment of frequencies leads to several more possible combinations: $H_u(1) + H_u(4) = 1142\,\mathrm{cm}^{-1}$, $T_{1g}(1) + T_{2g}(2) = 1136\,\mathrm{cm}^{-1}$, $G_g(2)+G_g(2) = 1134\,\mathrm{cm}^{-1}$, $T_{1g}(1) + G_g(2) = 1135\,\mathrm{cm}^{-1}$. This would explain the significant width of the $1140\,\mathrm{cm}^{-1}$ Raman peak and the observation by *Gallagher* et al. [2.112] of a shoulder at $1136\,\mathrm{cm}^{-1}$. There are many examples in the literature of second-order Raman peaks which under resonance conditions become stronger than first-order peaks, so we conclude that our assignment of the $1140\,\mathrm{cm}^{-1}$ band to a second-order combination is not inconsistent with the existing experimental evidence. Moreover, the INS spectrum shows a pronounced dip at this frequency [2.79]. Additional work is probably needed to pin down the assignment of $G_g(4)$, but we prefer the assignment in Table 2.4.

$T_{2u}(4)$, $T_{2u}(5)$ and $G_g(6)$

Our analysis of the isotopic shift of the 1470 cm^{-1} $A_g(2)$ Raman peak shows that the presence of a mode about $12\,\mathrm{cm}^{-1}$ higher than this is critical for the explanation of the Raman spectra [2.21, 2.22]. We placed the $G_g(6)$ mode at this position in view of the theoretical predictions for this mode and the observation of a clear Raman peak at this frequency, which is in fact one of the strongest silent mode signals in the Raman spectrum of C_{60}. As noted earlier, all other G_g modes are seen as weak lines in the Raman spectrum. It was suggested by *Love* et al. [2.87], however, that the peak at $1482\,\mathrm{cm}^{-1}$ corresponds to the $T_{2u}(5)$ mode, and our calculations suggest that this would also be consistent with the observed isotope shifts. The $T_{2u}(5)$ peak has been placed at $1540\,\mathrm{cm}^{-1}$ by *Schettino* et al. [2.72]. We prefer the value $1525\,\mathrm{cm}^{-1}$ because it agrees with a weak ir peak [2.68, 2.70, 2.73], but this peak can also be explained as a second-order combination band. The MARI INS data of *Coulombeau* et al. [2.79] shows two peaks in this region, which is consistent with the presence of $G_g(6)$ and $T_{2u}(5)$. However, the separation between the two INS peaks is $24\,\mathrm{cm}^{-1}$, whereas according to Table 2.4 this separation is $43\,\mathrm{cm}^{-1}$. This is one of the only two significant discrepancies between the

INS data and Table 2.4, the other being the position of the highest-energy peak, after rescaling the energies as discussed above. The agreement with the INS data could be improved by placing the T_{2u} peak at 1506 cm^{-1}, but it is uncertain whether the quality of the data guarantees this assignment.

2.3 Infrared Absorption Intensities of C_{60}

From (2.17), the area of the first-order ir absorption peak due to mode f is proportional to the square of the dipole moment derivative $\vec{\boldsymbol{\mu}}_f = [\partial\vec{\boldsymbol{\mu}}(\boldsymbol{u})/\partial d_f]_0$. This quantity is determined by the mode eigenvector $\chi(f)$ and the real-space effective charges $[\partial\vec{\boldsymbol{\mu}}(\boldsymbol{u})/\partial u(\ell\alpha)]_0$ through (2.15) [or its equivalent (2.18)]. Physically, $\vec{\boldsymbol{\mu}}_f$ is the dipole moment of the system (per unit displacement d_f) after displacing the real-space effective charges according to the mode pattern.

If the adiabatic many-electron ground state wave function $\phi_0(\boldsymbol{x},\boldsymbol{u})$ is known, the real-space effective charges can be computed directly from the displacement derivatives of (2.13), and the results combined with calculated mode eigenvectors to yield predicted ir strengths. Table 2.6 gives such electronic-structure-based predictions, for six representative models. The strengths are normalized to that for $T_{1u}(1)$, and for comparison, the last column gives the strength ratios measured by *Chase* et al. [2.84]. Columns two and three give results from empirical tight binding models [2.113, 2.114], the fourth column gives ratios from quantum chemistry semi-empirical MNDO calculations [2.43], and the results in columns five through seven are from three implementations of first-principles LDA techniques. These are the pseudoatomic orbital scheme [2.54], which we have previously used to compute the geometries and normal modes of a variety of fullerenes [2.53] (column five), the Car-Parrinello method [2.48] as used by *Bertsch* et al. [2.113] (column six), and density functional perturbation theory calculations of *Giannozzi* and *Baroni* [2.47] (column seven). Our choice to compare with the measured ir strength ratios of [2.84] is somewhat arbitrary, and for comparison, Table 2.7 lists ratios measured by several groups [2.84, 2.115, 2.116, 2.117, 2.118, 2.119]. All of these experiments were done on condensed phase samples of C_{60}, either thin film or crystalline, and it is not clear to what extent the differences between the various experiments are due to intermolecular effects, temperature variations, sample/substrate interactions, etc. Since the ir strengths provide a stringent test of theoretical models, it would be very useful to have low temperature ir data available for unperturbed icosahedral C_{60} in the gas phase.

The theoretical ir strength ratios in Table 2.6 exhibit a wide variation from model to model, much larger than the variations between the different measurements given in Table 2.7. The mode dipole moment derivatives $\vec{\boldsymbol{\mu}}_f$ describe vibrationally-induced electronic charge redistribution and offer a more difficult test of electronic-structure-based theoretical models than do

Table 2.6. Electronic-structure-based calculations of the relative absorption strengths (peak areas) for the four first-order ir active modes of icosahedral C_{60}. Results are given for three semi-empirical models (TBA and MNDO) and three first-principles models (LDA). The strengths are normalized to that of $T_{1u}(1)$. The last column gives measured values, for comparison

Mode	TBA[a]	TBA[b]	MNDO[c]	LDA[d]	LDA[e]	LDA[f]	Experiment[g]
$T_{1u}(1)$	1.0	1.0	1.0	1.0	1.0	1.0	1.0
$T_{1u}(2)$	0.47	0.08	0.33	0.65	0.65	0.63	0.34
$T_{1u}(3)$	0.17	0.18	1.9	1.4	0.59	0.36	0.28
$T_{1u}(4)$	2.7	1.7	4.4	0.31	0.41	0.57	0.34

[a] *Bertsch* et al. [2.113]
[b] *Esfarjani* et al. [2.114]
[c] *Stanton* and *Newton* [2.43]
[d] *Adams* et al. [2.53]
[e] *Bertsch* et al. [2.113]
[f] *Giannozzi* and *Baroni* [2.47]
[g] *Chase* et al. [2.84]

Table 2.7. Experimental first-order ir relative strengths. All of the measurements were on thin film samples, except for those of *Winkler* et al. [2.117], which were on single crystals. Where known, the temperature is listed

	Chase et al.[a]	*Fu* et al.[b]	*Martin* et al.[c]	*Winkler* et al[d] 100K	*Onoe* and *Takeuchi*[e] 300K	*Hara* et al.[f] 90K
$T_{1u}(1)$	1.0	1.0	1.0	1.0	1.0	1.0
$T_{1u}(2)$	0.34	0.39	0.48	0.28	0.30	0.27
$T_{1u}(3)$	0.28	0.29	0.45	0.16	0.22	0.20
$T_{1u}(4)$	0.34	0.36	0.38	0.36	0.24	0.38

[a] Ref. [2.84]
[b] Ref. [2.115]
[c] Ref. [2.116], as quoted in [2.17]
[d] Ref. [2.117]
[e] Ref. [2.118]
[f] Ref. [2.119]

predictions of either equilibrium geometries or normal modes. Moreover, errors in the $\vec{\mu}_f$'s are magnified when these quantities are squared to form the ir strengths. Although none of the electronic-structure-based calculations yield a highly quantitative reproduction of the measured ratios, the first-principles LDA-based results of columns five to seven are seen to be better qualitatively than those based on empirical electronic structure methods. This is particularly true for the ratios given in columns six and seven, which derive

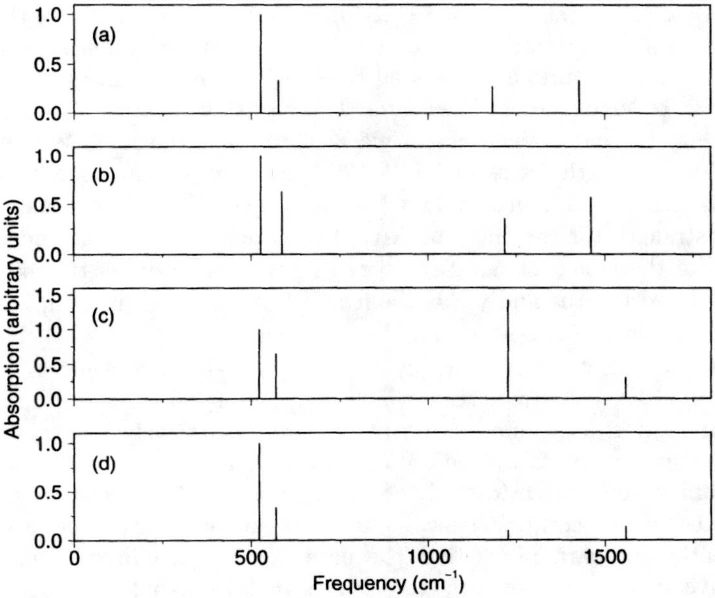

Fig. 2.12. Measured first-order ir spectra for C_{60}, together with three calculated spectra. The integrated peak intensities are given as vertical lines and are normalized to that for $T_{1u}(1)$. Panel (**a**) gives the experimental spectrum of *Chase* et al. [2.84], and panels (**b**) and (**c**) give the results of first-principles LDA-based theoretical predictions by *Giannozzi* and *Baroni* [2.47] and *Adams* et al. [2.53], respectively. The theoretical spectrum in panel (**d**) is based on our implementation of the two-parameter phenomenological model of *Fabian* [2.17]

from more exact LDA implementations than we used to obtain the results in column five. In Fig. 2.12, we compare both the predicted ir frequencies and relative strengths with the experimental values of *Chase* et al. [2.84], shown in the top panel. The first-principles LDA-based spectrum predicted by *Giannozzi* and *Baroni* [2.47] is shown in panel (b), and the qualitative agreement with the experimental data is seen to be quite good. Panels (c) and (d) show, respectively, the first-principles LDA-based spectrum of *Adams* et al. [2.53] and a phenomenological model fit to be discussed below.

The pseudoatomic orbitals LDA scheme [2.54] that underlies our calculations for panel (c) is convenient for a direct evaluation of the real-space effective charges, since it can be used for an arbitrary atomic configuration. Analogous to our calculation of the force constant matrix, we displace the atom at a given site ℓ successively in the x, y, and z directions by a small amount, while keeping the remaining atoms fixed at their equilibrium positions, and use (2.13) to calculate the dipole moment of the system, per unit displacement. Provided the displacements are kept small enough that the dipole moment remains linear in the displacements, this yields the 3×3 matrix of real-space effective charges $\{\mu_\beta(\ell\alpha) \equiv [\partial\mu_\beta(\boldsymbol{u})/\partial u(\ell\alpha)]_0 \approx \mu_\beta/u(\ell\alpha)\}$ associated with the site. In practice, this is done for both positive and negative displacements, and the results are averaged to eliminate the nonlinear contributions of even order in the displacements. The resulting effective charges are combined with the mode eigenvectors via (2.15) to give the $\vec{\boldsymbol{\mu}}_f$'s. While our first-principles strength ratios in Fig. 2.12(c) are seen to be in qualitative agreement with the measured values for the second and fourth peaks, the calculated strength for the third peak is almost an order of magnitude too large, despite the fact that our LDA-based method reproduces the experimental C_{60} bond lengths and yields the ir and Raman mode frequencies with an average deviation of better than 4%, as discussed in Sect. 2.1.4. It is interesting to note that when we apply the same first-principles method to the more complicated C_{70} molecule, which is of D_{5h} symmetry and has 31 symmetry-allowed ir active modes, our results are qualitatively improved with respect to experiment, as will be brought out below.

Given the difficulties of quantitatively predicting ir strengths directly from the electronic states, one can undertake to derive phenomenological models having sufficiently few parameters that the data can be meaningfully fit. If a quantitative model can be achieved, studies of isotope-induced spectral changes can be used to reveal information on spectroscopically inactive modes, as discussed in Sect. 2.2.5. Moreover, the vast number of possible fullerene structures and the similarity of their local bonding renders them attractive for attempts to develop phenomenological models which can be easily transferred from simpler to more complicated systems.

An ambitious model of this sort is that of *Sanguinetti* et al. [2.63], which combines the bond charge model of *Weber* [2.61] with a bond polarizability approach [2.120,2.121]. The result is a 6-parameter model which allows one to compute the equilibrium geometry, normal modes, ir strengths, and Raman strengths. The number of independent parameters is reduced to four when first-principles LDA-calculated equilibrium positions are input, and three of these are determined by fitting the frequencies of the 14 first-order ir and Raman modes. The fit reproduces these frequencies with an average deviation of 3.8%. The ir strengths can then be predicted (the Raman strengths require an additional parameter), yielding the strength ratios 1.0, 0.03, 0.4, 0.2. Unfortunately, while the calculated strength ratios for peaks three and four are in qualitative agreement with experiment, the negligible predicted intensity for $T_{1u}(2)$ is strongly at variance with the fact that this is the second most intense measured peak. Moreover, the result of transferring this model to C_{70} in [2.63] is a predicted ir spectrum which is dominated by a very strong peak near $550\,\mathrm{cm}^{-1}$, whereas the experimental spectrum is dominated by a very intense peak at $1430\,\mathrm{cm}^{-1}$. We do not know how well the situation might be improved by simply using the model to fit the C_{60} ir strengths, at the expense of the mode frequencies. The first-order Raman spectrum predicted by the model for C_{60} is better than for the ir, but the eight H_g peaks are consistently too strong. The predicted Raman spectrum for C_{70} is similarly better than for the ir, although the large number of symmetry-allowed modes (53, encompassing 3 irreducible representations) makes the peak assignments and detailed comparisons difficult.

Since there are many models that give good results for the vibrations, it is reasonable to focus on just the effective charges when trying to develop quantitative phenomenological models for the ir strengths, feeding in the mode eigenvectors from separate calculations. In contrast to bond charge models, which emphasize the effect of the electronic density between the atoms, a simplified classical model for the effective charges which focuses on the π-electron system has been developed by *Fabian* [2.17]. With the mode eigenvectors given, the result is a two-parameter model for the relative strengths of the first-order ir peaks.

The dipole moment of a given mode in Fabian's model is determined by how the atomic motion affects the classical coordinates $\{\vec{r}_i(t)\}$ of the 60 π electrons, each of which belongs to a parent carbon atom, at position $\vec{R}_i(t)$. Each carbon is bonded via two "single" bonds and a "double" bond to three neighbors, whose positions are denoted by $\vec{R}_{ij}(t)$, for $j = 1, 3$. Due to the interactions with the three adjacent bonds, the coordinate for π electron i is assumed to vary as $\vec{r}_i(t) = [1 + c_2\Delta_i(c_1, t)]\hat{n}_i(c_1, t)$, where c_1 and c_2 are the model's fitting parameters, $\hat{n}_i(c_1, t)$ is the unit normal vector to the plane defined by the three neighboring atoms' instantaneous positions (which are rescaled as described below), and $\Delta_i(c_1, t)$ is the difference between the instantaneous and equilibrium values of the distance of the parent atom i from this plane. The dipole moment is simply taken to be $\vec{\mu}(t) = C\sum_{i=1}^{60}\vec{r}_i(t)$, where the constant prefactor C is unimportant for relative intensities. The parameter c_1 accounts in a simplified way for the different interactions between electron i and its adjacent single or double bonds. This is done by rescaling the position of the doubly-bonded neighbor by a factor c_1: $\vec{R}_{id}(t) \to c_1\vec{R}_{id}(t)$; no rescaling is done when j represents either of the two singly-bonded neighbors. Geometrically, this rescaling causes the direction of the electron coordinate $\vec{r}_i(t)$ to deviate from the instantaneous normal to the actual plane of neighbors, by instead tilting away (for positive c_1) from the double bond. With just normal mode f excited, the atomic positions are $\vec{R}_i(t) = \vec{R}_{i,0} + \vec{\chi}(i|f)d_f(t)$, and the unit vectors become $\hat{n}_i(c_1, t) = \hat{n}_{i,0}(c_1) + \Delta\hat{n}_i(c_1, t)$, where the zero subscripts denote equilibrium configuration quantities. It is then straightforward to evaluate the mode dipole moment derivatives $\vec{\mu}_f = [\partial\vec{\mu}(u)/\partial d_f]_0$ as a function of the atomic equilibrium positions, the mode eigenvectors, and the model parameters c_1 and c_2.

Using our first-principles eigenvectors [2.53], and varying the parameters c_1 and c_2, we obtain the calculated relative intensities shown in the fourth panel of Fig. 2.12. Our parameter values are $(c_1, c_2) = (1.34, 0.67)$. These may be compared with the values (1.59, 0.67) obtained by *Fabian* [2.17], who used a different model for the eigenvectors and fit to the relative intensities measured by *Martin* et al. [2.116], whereas we are comparing with the data of *Chase* et al. [2.84]. Although our fit value for the intensity ratio of the first to the second peak is perfect, the third and fourth peaks are roughly a factor of 1/3 too low. Moreover, forcing a better fit for either or both of these two peaks worsens the overall fit. Given that the model has two free parameters for just three observed ir intensity ratios, the results again illustrate the very stringent theoretical test provided by the ir intensities.

The ir intensity model of Fabian is classical and depends only on the geometry of the vibrating molecule, with the central role played by the plane defined by the three neighbors of each carbon atom. To see if the model might have a useful transferability to more complicated trigonally-bonded fullerenes, we used it, with our above values of c_1 and c_2, to compute the relative intensities for D_{5h} C_{70} molecules. These have eight distinct bonds rather than two, and we made the the rough approximation of simply dividing the bonds into two sets, according to whether their lengths are larger or smaller than 1.425 Å, the midpoint between the measured C_{60} single and double bond lengths. Out of the 31 symmetry-allowed ir active peaks, only 9 are predicted to have intensities greater than 15% of the maximum. This plus the fact that these peaks mainly occur in two frequency regions ($\sim 400\,\mathrm{cm}^{-1}$–$800\,\mathrm{cm}^{-1}$, and $\sim 1200\,\mathrm{cm}^{-1}$–$1600\,\mathrm{cm}^{-1}$) are in qualitative agreement with experiment. However, the strongest predicted peaks occur in the first of these regions, whereas the experimental spectrum [2.122] is dominated by a very intense peak at $1430\,\mathrm{cm}^{-1}$, as noted earlier. Indeed, the overall predicted spectrum is roughly similar to the C_{70} spectrum predicted by *Sanguinetti* et al.'s much more complicated bond charge model [2.63]. It is interesting to note that our fully first-principles calculation of the ir strengths for C_{70} are in good qualitative agreement with experiment, being dominated by a very strong predicted peak at $1532\,\mathrm{cm}^{-1}$ [2.123].

In summary, quantitative theoretical predictions of the first-order ir strengths of C_{60} are challenging for both first-principles and phenomenological models. This is not surprising for the first-principles models, given the sensitivity of the intensities to the electronic states and their matrix elements. Regarding phenomenological approaches, it would be very helpful if a simple model could be developed which easily and reliably transfers between different fullerenes, but such a model does not yet exist. It is interesting that for the presumably more complicated problem of off-resonance Raman intensities, a bond-polarizability model based on a quantitative fit to the observed C_{60} Raman spectrum appears to transfer quite well to more complicated fullerenes, as will be brought out in the next section. These include polymeric fullerenes such as the C_{120} dimer, which include 4-coordinated as well as 3-coordinated atoms. The bond-polarizability model has a total of six parameters (three for each type of bond, single or double), of which five are needed to fit the measured intensity ratios for the ten first-order Raman active modes of C_{60}; thus the model parameters are well-determined by the data, and the question of transferability to other fullerenes can be addressed meaningfully. But for ir spectra, the high symmetry of C_{60} restricts the first-order active modes to only four, and this may be simply too few to pin down the parameters of a realistic and transferable model. Perhaps a better starting point for this purpose would be the richer ir spectrum of the C_{70} molecule.

2.4 Raman Intensities of C_{60}

Our discussion of Raman intensities begins in Sect. 2.4.1 with a detailed account of the relative intensities of the 10 first-order peaks observed under off-resonance conditions, where the incident photon energy is well below that of any electronic transition. Section 2.4.2 concerns the more difficult experimental problem of absolute Raman intensities. Resonance Raman scattering by C_{60}, for which there is still relatively little work, is discussed in Sect. 2.4.3.

2.4.1 Relative Intensities for Off-Resonance Scattering

As discussed in Sect. 2.1.2, the key quantity determining off-resonance Raman intensities is the system's electronic polarizability $\mathcal{P}_{\alpha\beta}(\omega_{\mathrm{L}}, \boldsymbol{u})$ for fixed atomic configuration $\boldsymbol{u}$ and incident light frequency ω_{L}, where the latter is assumed to be well below any electronic transition frequency. The static limit $\omega_{\mathrm{L}} = 0$ leads to the usual off-resonance selection rules, e.g., for icosahedral C_{60}, the first-order Raman activity is restricted to the A_g and H_g modes. From (2.24) and (2.25), the intensity of first-order scattering by mode f is proportional to the square of the mode's static electronic polarizability derivative

$$P_{\alpha\beta,f} = \sum_{\ell\gamma} \left[\frac{\partial \mathcal{P}_{\alpha\beta}(\boldsymbol{u})}{\partial u(\ell\gamma)} \right]_0 \chi(\ell\gamma|f). \qquad (2.36)$$

The dynamical properties enter this quantity through the normal mode eigenvectors $\boldsymbol{\chi}(f)$, while the real-space polarizability derivatives are determined by the electron-vibrational coupling. Section 2.1.4 reviewed theoretical calculations of C_{60} vibrations, and here we focus on computing the polarizability derivatives.

Just as for the ir effective charges and the dynamical properties, two general approaches can be undertaken to compute the polarizability derivatives $P_{\alpha\beta}(\ell\gamma)$, namely first-principles and phenomenological. For first-principles calculations directly from the basic Born–Oppenheimer electronic states, one could try to use the fundamental linear response theory result (2.20) for the electronic polarizability. Unfortunately, this would require knowledge of all the many-electron *excited* as well as ground states, and these are very difficult to calculate with current state-of-the-art first-principles techniques. To avoid this, one could instead undertake to solve the many-electron ground state problem in the presence of an applied electrostatic field, with the coupling treated within the dipole approximation, i.e., $-\vec{M}_e \cdot \vec{E}$, where $\vec{M}_e$ is the electronic dipole moment operator $\sum_e q_e \vec{x}_e$. A calculation of the change in the expectation value of $\vec{M}_e$, per unit applied field with the atoms fixed at their equilibrium positions would then yield the static electronic polarizability. To obtain the polarizability *derivatives* would involve the further step of solving the many-electron ground state problem in the presence of both a static E field and a displaced atomic configuration. This is equivalent to computing the *third* derivative of the many-electron ground state energy, twice with respect to an external static electric field and once with respect to an atomic displacement. LDA-based calculation of this sort have been made for small molecules [2.124, 2.125]. *Giannozzi* and *Baroni* [2.47] have used density functional perturbation theory to carry out such calculations for the polarizability derivatives of C_{60}. The resulting Raman peak intensities were not compared with experiment in [2.47], but they agree quite well with measured first-order off-resonance relative intensities – the theoretical/experimental comparison is shown in Fig. 3 of [2.126].

Electronic polarizability derivative calculations push first-principles techniques to a more difficult level than do purely vibrational or structural calculations and can become very intensive numerically. Accordingly, phenomenological models are useful. In the previous section we discussed the model of *Sanguinetti* et al. [2.63], for computing the normal modes, and ir and Raman spectra of C_{60} and C_{70}. In 1993, *Snoke* and *Cardona* [2.91] combined a simple bond-polarizability model with separate calculations of the vibrations to compute the relative intensities of the first-order Raman lines of C_{60}. We have found this approach to be convenient for studies of isotope effects, as discussed in Sect. 2.2.5, and for predicting the Raman spectra of more complex structures, such as polymerized fullerenes [2.53].

The bond polarizability approach was developed for molecules [2.127], but it is also useful for solid-state applications, as is reviewed in [2.121]. The basic assumption is that the static electronic polarizability of a covalent system can be written as a sum of contributions from the individual bonds. In its simplest form, the contribution $\Pi_{\alpha\beta}(\vec{R})$ from the bond associated with a pair of atoms separated by $\vec{R}$ is taken to be a cylindrically symmetric function of the bond length. Thus, for a bond parallel to the x-axis, the only nonzero elements of the bond polarizability are $\Pi_{xx}(R) \equiv \alpha_{\parallel}(R)$ and $\Pi_{yy}(R) = \Pi_{zz}(R) \equiv \alpha_{\perp}(R)$. For a general orientation, this becomes

$$\Pi_{\alpha\beta}(\vec{R}) = \frac{R_\alpha R_\beta}{R^2}\alpha_{\parallel}(R) + \left(\delta_{\alpha\beta} - \frac{R_\alpha R_\beta}{R^2}\right)\alpha_{\perp}(R). \tag{2.37}$$

The bond polarizability derivatives are then

$$\frac{\partial \Pi_{\alpha\beta}(\vec{R})}{\partial R_\gamma} = \left[\frac{\alpha_{\parallel}(R) - \alpha_{\perp}(R)}{R}\right](\hat{R}_\alpha\delta_{\beta\gamma} + \hat{R}_\beta\delta_{\alpha\gamma} - 2\hat{R}_\alpha\hat{R}_\beta\hat{R}_\gamma)$$
$$+\{\alpha'_{\perp}(R)\delta_{\alpha\beta} + [\alpha'_{\parallel}(R) - \alpha'_{\perp}(R)]\hat{R}_\alpha\hat{R}_\beta\}\hat{R}_\gamma, \tag{2.38}$$

where $\hat{\boldsymbol{R}}$ is a unit vector along $\vec{\boldsymbol{R}}$.

For (2.36), we need the derivatives of the system's polarizability with respect to the atomic displacements $\{u(\ell\gamma)\}$, evaluated at the equilibrium configuration. Giving the atom at site ℓ a displacement $\vec{\boldsymbol{u}}(\ell)$ while keeping the remaining atoms fixed, we have $\vec{\boldsymbol{R}}(\ell B) = \vec{\boldsymbol{R}}_0(\ell B) - \vec{\boldsymbol{u}}(\ell)$, where $\vec{\boldsymbol{R}}(\ell B)$ is the bond vector from atom ℓ to any one of its bonded neighbors ℓB, and $\vec{\boldsymbol{R}}_0(\ell B)$ is the corresponding equilibrium configuration bond vector. Then

$$\left[\frac{\partial \mathcal{P}_{\alpha\beta}(\boldsymbol{u})}{\partial u(\ell\gamma)}\right]_0 = -\sum_{B}{}' \left[\frac{\partial \Pi_{\alpha\beta}[\vec{\boldsymbol{R}}(\ell B)]}{\partial R_\gamma(\ell B)}\right]_0 , \tag{2.39}$$

where the prime over the sum signifies that only the bonds attached to site ℓ are summed. Combining (2.36), (2.38) and (2.39), we can write the mode polarizability derivatives in the convenient form [2.22]

$$\begin{aligned} P_{\alpha\beta,f} = -\textstyle\sum_\ell \sum_B' \Bigg[& \left\{\frac{\alpha'_\parallel[R_0(\ell B)] + 2\alpha'_\perp[R_0(\ell B)]}{3}\right\} \hat{\boldsymbol{R}}_0(\ell B)\cdot\vec{\chi}(\ell|f)\delta_{\alpha\beta} \\ & + \left\{\frac{\alpha'_\parallel[R_0(\ell B)] - \alpha'_\perp[R_0(\ell B)]}{3}\right\} \left[3\hat{R}_{0\alpha}(\ell B)\hat{R}_{0\beta}(\ell B) - \delta_{\alpha\beta}\right] \hat{\boldsymbol{R}}_0(\ell B)\cdot\vec{\chi}(\ell|f) \\ & + \left\{\frac{\alpha_\parallel[R_0(\ell B)] - \alpha_\perp[R_0(\ell B)]}{R_0(\ell B)}\right\} \Big[\hat{R}_{0\alpha}(\ell B)\chi(\ell\beta|f) + \hat{R}_{0\beta}(\ell B)\chi(\ell\alpha|f) \\ & - 2\hat{R}_{0\alpha}(\ell B)\hat{R}_{0\beta}(\ell B)\hat{\boldsymbol{R}}_0(\ell B)\cdot\vec{\chi}(\ell|f)\Big]\Bigg] . \end{aligned} \tag{2.40}$$

The first two terms give the bond-stretching induced changes of the isotropic and anisotropic parts of the bond polarizabilities, respectively, while the third term gives the bond-rotation induced changes of the anisotropic parts of the polarizabilities. For C_{60}, the first term contributes only to the scattering by the two A_g modes, and the other two terms contribute only to the scattering by the eight H_g modes. On a practical note, the above approach is convenient in that it uses the mode eigenvectors in just a single sum over sites, avoiding the need to compute their differences, i.e., the "bond displacements."

For C_{60}, the bond sum for each site extends over two "single" bonds and one "double" bond, so that there are six independent polarizability parameters, namely $\alpha'_\parallel$, $\alpha'_\perp$, and $\alpha_\parallel - \alpha_\perp$, for each of the two bond types. For the relative intensities of the Raman peaks, only 5 parameters are needed, and we have fit *Chase* et al.'s [2.84] off-resonance Raman spectra,[5] normalizing the parameters with respect to $(\alpha_\parallel - \alpha_\perp)/R_0$ for single bonds. Our fit ratios are

[5] Note that there is a discrepancy between the peak heights quoted in Table III of [2.84] and the peak heights shown in Fig. 7 of the same paper. Hence, we obtained the experimental peak areas as follows. First we computed Lorentzian areas by combining the peak heights shown in the experimental spectrum in Fig. 7 of [2.84] with the peak widths listed in that paper's Table III. We then averaged the resulting areas with those obtained from a second Raman spectrum,

Table 2.8. Parameter ratios for our bond-polarizability model fit to the first-order C_{60} Raman spectrum measured by *Chase* et al. [2.84] (see footnote 5). The incident light frequency used in the experiments was $13289\,\mathrm{cm}^{-1}$, below resonance with electronic transitions. The ratios are taken with respect to the quantity $N \equiv [\alpha_{\|}(S) - \alpha_{\perp}(S)]/R_0(S)$, where S denotes single bonds. Our calculated C_{60} bond lengths are $R_0(S) = 1.45$Å and $R_0(D) = 1.40$Å for single and double bonds, respectively, in agreement with the experimental values. The fit spectra are shown in Fig. 2.13

Ratio	Fit value
$[\alpha'_{\|}(S) - \alpha'_{\perp}(S)]/N$	2.80
$[\alpha'_{\|}(S) + 2\alpha'_{\perp}(S)]/N$	2.73
$[\alpha'_{\|}(D) - \alpha'_{\perp}(D)]/N$	3.18
$[\alpha'_{\|}(D) + 2\alpha'_{\perp}(D)]/N$	8.98
$[\alpha_{\|}(D) - \alpha_{\perp}(D)]/[NR_0(S)]$	0.256

listed in Table 2.8, and Fig. 2.13 gives our calculated spectra. The top panel of the figure shows the measured frequencies and intensities, and the other two panels give the freqencies and intensities computed using the parameters of Table 2.8 and two different sets of first-principles normal mode frequencies and eigenvectors. For panel (b), we used the eigenfrequencies and eigenvectors of *Giannozzi* and *Baroni* [2.47], whereas for panel (c) we used those of *Adams* et al. [2.53]. The actual fitting of the bond polarizability parameters was done using the eigenmodes of [2.47], but the computed intensities in panels (b) and (c) are seen to be quite similar.

Similar to the ir case, our use of Chase et al.'s experimental off-resonance spectra is somewhat arbitrary, and in Table 2.9 we compare their intensities with those measured by three other groups. All four experiments were done using 1064 nm incident excitation. The table lists peak heights, since the areas were not reported in every case. All four groups observed a sideband at $266\,\mathrm{cm}^{-1}$, just below the $H_g(1)$ peak. In the data of *Chase* et al. [2.84], this was reported as a shoulder, but in the other three spectra a separate peak was resolved, whose height is also listed in the table. The shoulder was not included in fitting the data of Chase et al. to obtain the calculated spectrum of Fig. 2.13b.

The above bond polarizability approach was discussed in [2.126], where we explored the extent to which polarizability parameters for small hydrocarbon molecules, namely ethane for the C–C single bonds and ethylene for the double bonds, can be transferred to C_{60} and C_{70}. As noted in Sect. 2.3, C_{70} is of D_{5h} symmetry and has eight distinct bonds, and we thus made the rough approximation of dividing the bonds into two sets, according to

kindly sent to us by B. Chase. The averaged peak strengths are shown here as the experimental intensities in Fig. 2.13, and we obtained our bond polarizability model parameters of Table 2.8 by fitting these intensities.

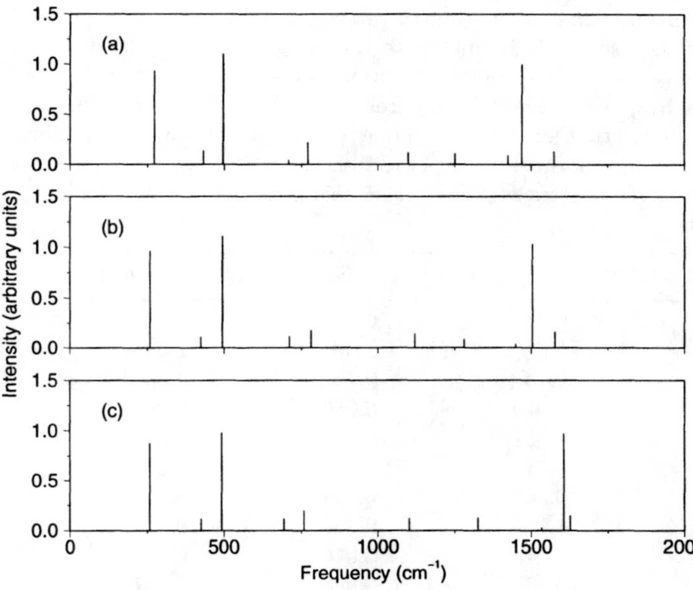

Fig. 2.13. Measured off-resonance first-order Raman spectra for C_{60}, together with two calculated spectra. The integrated intensities are given as vertical lines. The experimental spectrum of panel (**a**) is that of *Chase* et al. [2.84] (see footnote 5), measured at $\omega_L = 13289\,\mathrm{cm}^{-1}$. The spectrum of panel (**b**) was computed using (2.25) and (2.40) together with the first-principles normal mode eigenfrequencies and eigenvectors of *Giannozzi* and *Baroni* [2.47], with the bond polarizability parameters adjusted to fit the measured spectrum of panel (**a**). The resulting parameter values are given in Table 2.8. For panel (**c**), we used these same parameter values, together with the normal mode eigenfrequencies and eigenvectors from the first-principles calculations of *Adams* et al. [2.53] While absolute intensities were not computed, the vertical scales for panels (**b**) and (**c**) are the same, in order to show the effect of the two different sets of mode frequencies and eigenvectors

whether their bond lengths are larger or smaller than 1.425 Å, the midpoint between the observed C_{60} single and double bond lengths. The single (double) bond polarizability parameters were then used for bonds in the first (second) set. The use of static hydrocarbon polarizabilities was found to give very good agreement with the measured static polarizabilities for both C_{60} and C_{70}. For the more complicated case of Raman scattering, which is sensitive to the derivatives of the polarizability, the use of hydrocarbon parameters led to only fair agreement with the observed relative intensities for C_{60}, with significant discrepancies occurring for modes above $1000\,\mathrm{cm}^{-1}$. Simply fitting the observed C_{60} Raman intensities yielded the very good agreement shown here in Fig. 2.13. Interestingly, for the C_{70} Raman intensities, we found that the use of either the hydrocarbon parameters or the C_{60} fit values yielded comparable qualitatively good agreement with experiment, much better than

Table 2.9. Experimental first-order Raman intensities, obtained using incident light from a Nd:YAG laser at 1064 nm. Since the integrated strengths were not reported in each case, but the spectral resolutions were similar, the table gives the measured peak heights obtained directly from the published spectra. The peak heights are normalized to that for the $A_g(2)$ mode. For the $H_g(1)$ peak, the numbers in parentheses give the height of a second peak, observed at 266 cm^{-1} in each of the measured spectra. In the spectra of *Chase et al.* this peak appears as a shoulder, so no height is listed

	Chase et al.[a]	*Dennis* et al.[b]	*Birkett* et al.[c]	*Lynch* et al.[d]
$A_g(1)$	1.1	0.77	0.65	0.80
$A_g(2)$	1.0	1.0	1.0	1.0
$H_g(1)$	0.91	0.83 (0.21)	0.72 (0.19)	0.96 (0.24)
$H_g(2)$	0.076	0.040	0.056	0.045
$H_g(3)$	0.014	0.011	0.016	0.013
$H_g(4)$	0.18	0.11	0.11	0.13
$H_g(5)$	0.049	0.045	0.035	0.046
$H_g(6)$	0.056	0.046	0.050	0.048
$H_g(7)$	0.013	0.014	0.016	0.017
$H_g(8)$	0.087	0.072	0.098	0.087

[a] Ref. [2.84]; see footnote 5
[b] Ref. [2.128]
[c] Ref. [2.129]
[d] Ref. [2.130]

the hydrocarbon set gave for the C_{60} intensities. It should be noted, though, that the situation for C_{70} is made somewhat difficult by the lack of unambiguous identification of the symmetry types of several of the observed peaks. Nevertheless, given the crudeness of our bond-partitioning scheme described above, the results are encouraging for the transfer of polarizability parameters to more complicated fullerenes. In the papers cited in [2.53], we have used the same bond partitioning scheme (again with the partitioning bond length of 1.425 Å), together with the fit C_{60} bond polarizability parameters of Table 2.8, to compute the first-order Raman intensities for a variety of polymerized fullerenes. These systems typically contain hundreds of atoms, a few of which are tetragonally rather than trigonally bonded, and they have low symmetry, such that the Raman spectra are quite congested. The use of the fit C_{60} polarizability parameters to predict the Raman spectra of these systems seems a reasonable first approximation, since the spectral congestion is likely to wash out the detailed contributions from individual bonds. We have recently applied this model to new Raman data for the odd-numbered dimeric fullerene C_{119} and have found that it discriminates convincingly between different isomeric configurations [2.131].

Snoke et al. [2.132] have likewise studied the question of the transferability of hydrocarbon bond polarizability parameters for computing the first-order off-resonance Raman intensities of C_{60}, using two different force constant models for the normal modes. They also did a pure fit to the observed relative intensities, and some of their fit values of the polarizability parameters differ from those we obtain (Table 2.8), particularly for $[\alpha_{\|}(D)-\alpha_{\perp}(D)]/[\alpha_{\|}(S)-\alpha_{\perp}(S)]$, where the discrepancy is an order of magnitude. The origin of this difference is not known.

2.4.2 Absolute Raman Cross Sections

Lorentzen et al. [2.133] have measured the Raman cross-section for the $A_g(2)$ mode (pentagonal pinch) of C_{60}. For an excitation wavelength of 752 nm, they find $\mathrm{d}\sigma/\mathrm{d}\Omega = (2.09 \pm 0.29) \times 10^{-29}$ $\mathrm{cm}^2/\mathrm{sr}$. The measurements were performed on a 4×10^{-4} M solution of C_{60} in CS_2. In such a dilute solution it is reasonable to neglect the effects of C_{60}–C_{60} interactions, so that the measured cross-section can be related to the molecular polarizability derivative occurring in (2.25). However, we recall that this equation contains no local field corrections, whereas for for C_{60} in CS_2 these corrections should be included. Lorentzen et al. disposed of the local field corrrection by measuring the intensity of the $A_g(2)$ mode relative to that of the the $656\,\mathrm{cm}^{-1}$ CS_2 mode and by making the assumption that the local field correction is the same for the two modes. The cross-section for the CS_2 mode was obtained from independent measurements of CS_2/C_6H_6 solutions, taking advantage of the well-known absolute cross-section of the $992\,\mathrm{cm}^{-1}$ Raman mode of benzene [2.134]. Again, the local field corrections were assumed to drop out when taking ratios.

Since the primary motivation of *Lorentzen* et al.'s work [2.133] was to compare the experimental cross-section with predictions for the off-resonance case, they used the 752 nm line of a krypton laser instead of the more widely used green-blue lines of an argon laser. One could go even further below resonance by using the 1064 nm line of a Nd:YAG laser. However, as detailed in [2.133], CCD detectors have marginal performance in this range. Moreover, there are no well-calibrated absolute Raman intensities available at 1064 nm, whereas at 752 nm it is reasonable to extrapolate the results of *Schomaker* et al. [2.134] for the $992\,\mathrm{cm}^{-1}$ Raman peak of benzene, whose cross-section was determined for excitation wavelengths spanning the visible frequency range down to 656 nm. In this way Lorentzen et al. could avoid having to make a much more difficult *direct* measurement, as would be required for 1064 nm excitation. Thus the choice of 752 nm excitation appears to provide a reasonable compromise between cross-section accuracy and the off-resonance requirement. Using (2.25) to re-express the measured Raman cross-section in terms of the mode polarizability derivative, one obtains $P_{\alpha\alpha,A_g(2)} = [\partial P_{\alpha\alpha}/\partial d_{A_g(2)}]_0 = (5.66 \pm 0.40) \times 10^{-4}\,\mathrm{cm}^2\ \mathrm{g}^{-1/2}$. Assuming that the experimental conditions are well below important resonance effects, this value can be compared with predictions of the off-resonance theory, using either the bond polarizability model or *ab initio* calculations.

Bond polarizability models are particularly useful when the polarizability parameters are transferable among different systems, and absolute cross-section measurements provide an additional test of this property. If the transferability hypothesis is approximately correct for fullerenes, as is suggested by the comparison between the measured relative intensities of the first-order Raman peaks and the relative intensities we predict using hydrocarbon polarizability parameters (Fig. 1 in [2.126]), the error in the calculated absolute cross-section should be comparable to the error in the relative intensities. This is what *Lorentzen* et al. find [2.133]. When hydrocarbon polarizability parameters are used, (2.40) yields $P_{\alpha\alpha,A_g(2)} = 1.7 \times 10^{-4}\,\mathrm{cm}^2\ \mathrm{g}^{-1/2}$, which is within a factor of 3 of the experimental value.

First-principles methods are expected to make more accurate Raman intensity predictions than simple bond-polarizability models based on the transferability hypothesis. The off-resonance Raman intensities for diamond, silicon and germanium have been calculated by *Windl* [2.135], using a first-principles approach within the LDA. Excellent agreement with experiment was found for these three materials. As noted earlier, LDA-based methods for the prediction of infrared and Raman activities in small molecules have been investigated in [2.124] and [2.125], and for C_{60}, LDA-based calculations of the ir and Raman intensities have been published by *Bertsch* et al. [2.113] and by *Giannozzi* and *Baroni* [2.47]. The latter authors' predicted relative intensities for the 10 first-order Raman modes in C_{60} are shown in Fig. 2.13 and are in very good agreement with experiment. However, the comparison between theoretical and experimental *absolute* cross-sections is not completely straightforward. The calculations of [2.47] are based on plane wave expansions appropriate to an infinite fcc lattice with periodic boundary conditions, and to focus on the intramolecular properties, an artificially large lattice constant of 16 Å is used. Accordingly, the calculated quantity within this scheme is not the isolated molecular polarizability derivative but the derivative of the "crystalline" dielectric tensor relative to the vibrational normal coordinate. Assuming that the two quantities are related by a Clausius–Mossotti relation, one obtains $P_{\alpha\alpha,A_g(2)} = 2.7 \times 10^{-4}\,\mathrm{cm}^2\ \mathrm{g}^{-1/2}$, in reasonable agreement with the experimental results [2.133].

2.4.3 Resonance Raman Scattering

When the energy of the incident laser photons is well below the lowest absorption edge, the first-order Raman intensities are proportional to the derivatives of the static electronic polarizability with respect to the normal mode displacements. As discussed in Sect. 2.4.1, these can be computed using either state of the art first principles techniques [2.47, 2.53] or phenomenological methods [2.63,2.91]. However, when the laser excitation energy becomes comparable with that of the electronic absorption bands, the static polarizability

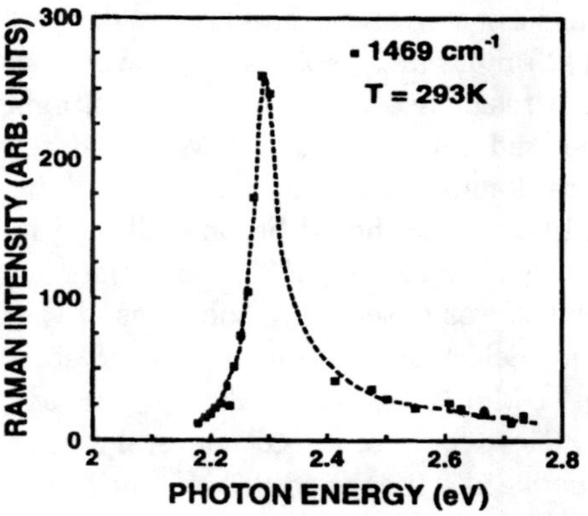

Fig. 2.14. Raman excitation profile for the $A_g(2)$ mode of C_{60}, from [2.138] The sample is a polycrystalline film of chromatographically purified material

approach breaks down. Under these conditions, the individual mode Raman intensities as a function of the exciting laser frequency (Raman excitation profiles) display strong resonance effects [2.19]. The study of these effects can provide detailed information on the electron–vibration coupling involving specific electronic states and the Raman active modes. Such information is important, for instance, in understanding the superconductivity mechanisms in doped C_{60}.

There are very few detailed experimental studies of Raman excitation profiles in C_{60}, and none of them cover the region between 3 eV and 5 eV, which includes the strongest electronic absorption bands. *Sinha* et al. [2.136] carried out resonance Raman experiments in C_{60} films using laser lines in the visible. At room temperature, they found a maximum at 2.4 eV in the Raman excitation profile for the $A_g(2)$ mode. Similar results were obtained by *Matus* et al. [2.137], but the profile maximum was found to occur at about 2.7 eV. These two experiments were performed on films which contained significant amounts of C_{70}. The existence of a Raman resonance in pure C_{60} films was confirmed by *Guha* et al. [2.138]. Figure 2.14 shows the results, which indicate a sharp maximum at 2.3 eV. *Denisov* et al. [2.139] also report an enhancement of the intensity of the $A_g(2)$ Raman peak, under 488 nm (2.54 eV) excitation from an argon laser. *Gallagher* et al. [2.140] compared Raman spectra obtained with the 407 nm and 413 nm lines of a krypton laser and analyzed them in terms of the so-called "D-term" scattering effects, which arise from non-adiabatic corrections. A few papers have also discussed surface-enhanced Raman scattering in C_{60} [2.141, 2.142, 2.143] and the effect of solvents on the Raman intensities [2.111].

Two explanations have been offered for the resonance in the excitation profile of the $A_g(2)$ mode in C_{60}. *Sinha* et al. [2.136] argued that the energy of the resonance was too low to be assigned to optically allowed transitions. Detailed studies of C_{60} in solution show that the first allowed transitions correspond to a broad structure with a maximum at 3.7 eV. The onset of this structure displays a very weak peak at 3.0 eV [2.144]. According to theoretical calculations [2.145], the first allowed transitions involve excited states derived from the h_u (HOMO) and t_{1g} (LUMO+1) molecular orbitals. Thus *Sinha* et al. [2.136] assigned the observed Raman resonance at 2.4 eV to transitions between crystalline electronic bands derived from the h_u (HOMO) and t_{1u} (LUMO) molecular orbitals. These transitions are optically forbidden in the icosahedral molecule, but it was argued that they should become allowed in the solid state. The mechanism proposed by *Sinha* et al. [2.136] was supported by the fact that no resonance enhancement was observed in solutions of C_{60} in CS_2. Theoretical calculations of the dielectric function of C_{60} crystals confirmed that "forbidden" HOMO $\rightarrow$ LUMO transitions make a significant contribution to the optical absorption at low energies [2.146]. Several strong transitions were identified in a range between 0.2 eV and 0.7 eV above the absorption threshold, which was found at 1.46 eV. When the experimental value of 1.72 eV [2.147, 2.148] is used, these strongest "forbidden" transitions are predicted to occur between 1.9 eV and 2.4 eV, overlapping the energy range within which the Raman resonance was observed by *Sinha* et al. [2.136] and *Guha* et al. [2.138]

An alternative explanation for the visible Raman resonance was proposed by *Matus* et al. [2.137]. These investigators suggested an interpretation based on the "A-term" in *Albrecht*'s molecular theory of resonance Raman scattering [2.149], with the resonant optical transition occurring between bands derived from the h_u and t_{1g} molecular orbitals. The transition energy between these two levels was placed at 2.7 eV. This is considerably lower than the value of 3.0 eV found for the lowest allowed transition in C_{60} in solution [2.144], although it is consistent with most absorption and ellipsometric measurements on C_{60} films and single crystals [2.147, 2.148]. The reason for the discrepancy is now well understood in terms of the width of the electronic bands derived from the molecular orbitals. For example, *Laouini* et al. [2.150] estimate a separation of 2.9 eV between the h_u (HOMO) and t_{1g} (LUMO+1) levels in the isolated molecule. When the solid is formed, the minimum separation between the energy bands derived from these molecular orbitals is reduced to 2.2 eV. This is consistent with an absorption peak at 3.7 eV (molecule) and 2.7 eV (solid).

Which of these two explanations for the $A_g(2)$ Raman excitation profile is correct remains undecided. *Guha* et al. [2.138] noted that the Raman resonance could be "turned on" in C_{60} dissolved in CS_2 simply by freezing the solution, suggesting that the Raman resonance is produced by symmetry breaking and not by the particular energy shift of the allowed transitions which occur in the solid state. One could argue, however, that the C_{60} molecule in the frozen solution suffers a perturbation which also lowers the energy of the HOMO $\rightarrow$ LUMO+1 transition, so that *Guha* et al.'s observation [2.138] is not necessarily inconsistent with *Matus*'s interpretation [2.137]. One should also keep in mind the possibility that both mechanisms may play a role of comparable significance. Unfortunately, the uncertainty as to the relevant electronic states makes it very difficult to analyze the excitation profiles in terms of different theoretical expressions. It is also apparent that deviations from the properties of the isolated molecule, which in the case of the vibrational structure are known to be small, play a key role in determining the measured Raman excitation profiles. This is not surprising when one notes that the Raman studies to date fail to cover the energy range between 3 eV and 5 eV, where the strongest electronic transitions occur and where one might expect a closer similarity between results for isolated molecules and for solid phases.

2.5 Conclusion

We have presented a survey of the vibrational properties of C_{60} and spectroscopic techniques used to investigate these properties, with the primary emphasis on ir absorption and Raman scattering. It is clear that great progress has been achieved in understanding this unique molecule. The level of detail with which the complex vibrational structure and the spectroscopic intensities are known in C_{60} is unmatched among fullerenes and nanotubes. However, significant gaps remain in this knowledge: some of the silent mode frequencies remain poorly known; the interplay between crystal field effects and isotopic perturbations produces complicated Raman and infrared spectra for which there are often no satisfactory explanations; little evidence exists on the phonon energy bands derived from the intramolecular modes; and while the infrared and off-resonance Raman intensities are reasonably well understood in terms of first-principles and phenomenological models, the information on resonant Raman excitation profiles remains preliminary. We hope that this review will stimulate interest in these topics and further enhance the role of C_{60} as the benchmark system for fullerene studies.

Acknowledgements

We would like to thank our collaborators Gary Adams, Stefano Baroni, Paolo Giannozzi, Suchismita Guha, Otto Sankey, and Kislay Sinha, each of whom played key roles in several developments discussed in this chapter. We also benefited from interactions with many colleagues who provided illuminating explanations, unpublished data, and many rounds of useful e-mail discussions. Among these, we would like especially to mention Manuel Cardona, Christian Coulombeau, Bruce Chase, Rolf Heid, Paul Heiney, Herve Jobic, Fabrizia Negri, Dan Neumann, Jun Onoe, and David Snoke. Our work has been supported by the National Science Foundation, under grants DMR 9624102 and DMR 9510182.

References

2.1 H. W. Kroto, J. R. Heath, S. C. O'Brien, R. F. Curl, R. E. Smalley: Nature **318**, 162 (1985)
2.2 W. Krätschmer, L. D. Lamb, K. Fostiropoulos, D. Huffman: Nature **347**, 354 (1990)
2.3 R. C. Haddon, A. F. Hebard, M. J. Rosseinsky, D. W. Murphy, S. J. Duclos, K. B. Lyons, B. Miller, J. M. Rosamilia, R. M. Fleming, A. R. Kortan, S. H. Glarun, A. V. Makhija, A. J. Muller, R. H. Eick, S. M. Zahurak, R. Tycko, G. Dabbagh, F. A. Thiel: Nature **350**, 320 (1991)
2.4 S. Ijima: Nature **354**, 56 (1991)
2.5 A. M. Rao, P. Zhou, K.-A. Wang, G. T. Hager, J. M. Holden, Y. Wang, W. T. Lee, X. X. Bi, P. C. Eklund, D. S. Cornet, M. A. Duncan, I. J. Amster: Science **259**, 955 (1993)
2.6 P. C. Eklund, A. M. Rao (eds.): *Fullerene Polymers and Fullerene Polymer Composites*, Springer Ser. Mates. Sci. **38** (Springer, Berlin, Heidelberg 1999) in press
2.7 W. E. Billups, M. A. Ciufolini: *Buckminster Fullerenes* (Wiley, New York 1993), p. 339
2.8 D. Koruga, S. Hameroff, J. Withers, R. Loutfy: *Fullerene C_{60}: History, Physics, Nanobiology, Nanotechnology* (Elsevier, Amsterdam 1993)
2.9 H. Ehrenreich, F. Spaepen (eds.): *Solid State Physics* (Academic, San Diego 1994) Vol. 48
2.10 T. Braun, A. Schubert, H. MacZelka, L. Vasvari: *Fullerene Research 1985–1993: A Computer-Generated Cross-Indexed Bibliography of the Journal Literature* (World Scientific, Singapore 1995)
2.11 M. S. Dresselhaus, G. Dresselhaus, P. C. Eklund: *Science of Fullerenes and Carbon Nanotubes* (Academic, San Diego 1996)
2.12 W. Kroto, D. R. M. Walton: *The Fullerenes: New Horizons for Chemistry, Physics, and Astrophysics of Carbon* (Cambridge Univ. Press, London 1997) p. 154
2.13 H. Aldersey-Williams: *The Most Beautiful Molecule: The Discovery of the Buckyball* (Wiley, New York 1995)
2.14 M. S. Dresselhaus, G. Dresselhaus, P. C. Eklund: J. Raman Spectrosc. **27**, 351 (1996)
2.15 P. C. Eklund, P. Zhou, K.-A. Wang, G. Dresselhaus, M. S. Dresselhaus: J. Phys. Chem. Solids **53**, 1391 (1992)
2.16 D. E. Weeks, W. G. Harter: J. Chem. Phys. **90**, 4744 (1989)
2.17 J. Fabian: Phys. Rev. B **53**, 13864 (1996)
2.18 M. Tinkham: *Group Theory and Quantum Mechanics* (McGraw Hill, New York 1964)
2.19 See, for instance, J. B. Page: in *Light Scattering in Solids VI*, ed. by M. Cardona, G. Güntherodt, Topics Appl. Phys. **68** (Springer, Berlin, Heidelberg 1991)
2.20 G. Placzek: *The Rayleigh and Raman Scattering*, UCRL Translation No. 526(L), from *Handbuch der Radiologie* VI 2 (Akademische Verlagsgs., Leipzig 1934) pp. 209–374
2.21 S. Guha, J. Menéndez, J. B. Page, G. B. Adams, G. S. Spencer, J. P. Lehman, P. Giannozzi, S. Baroni: Phys. Rev. Lett. **72**, 3359 (1994)

2.22 S. Guha, J. Menéndez, J. B. Page, G. B. Adams: Phys. Rev. B **56**, 15431 (1997)
2.23 J. D. Axe, S. C. Moss, D. A. Neumann: In *Solid State Physics*, ed. by H. Ehrenreich, F. Spaepen (Academic, New York 1994) Vol. 48, p. 149
2.24 J. E. Fischer, P. A. Heiney: J. Phys. Chem. Solids **54**, 1725 (1993)
2.25 J. M. Hawkins, A. Mayer, S. Loren, R. Nunlist: J. Am. Chem. Soc. **113**, 9394 (1991)
2.26 J. Menéndez, J. B. Page, S. Guha: Phil. Mag. B **70**, 651 (1994)
2.27 Q. Ying, J. Marecek, B. Chu: J. Chem. Phys. **104**, 2665 (1994)
2.28 P. Launois, S. Ravy, R. Moret: Phys. Rev. B **55**, 2651 (1997)
2.29 R. M. Fleming, T. Siegrist, P. M. Marsh, B. Hessen, A. R. Kortan, D. W. Murphy, R. C. Haddon, R. Tycko, G. Dabbagh, A. M. Mujsce, M. L. Kaplan, S. M. Zahurak: In *Mater. Res. Soc. Symp.* **206**, ed. by R. S. Averback, D. L. Nelson, J. Bernholc (Materials Research Society, Boston 1990)
2.30 P. A. Heiney: J. Phys. Chem. Solids **53**, 1333 (1992)
2.31 A. B. Harris, R. Sachidanandam: Phys. Rev. B **46**, 4944 (1992)
2.32 K. Rapcewicz, J. Przystawa: Phys. Rev. B **49**, 13193 (1994)
2.33 G. Burns, A. M. Glazer: *Space Groups for Solid State Scientists* (Academic, New York 1978) p. 200
2.34 M. C. Martin, J. Fabian, J. Godard, P. Bernier, J. M. Lambert, L. Mihaly: Phys. Rev. B **51**, 2844 (1995)
2.35 P. C. Chow, X. Jiang, G. Reiter, P. Wochner, S. C. Moss, J. D. Axe, J. C. Hanson, R. K. McMullan, R. L. Meng, C. W. Chu: Phys. Rev. Lett. **69**, 2943 (1992)
2.36 S. Ravy, P. Launois, R. Moret: Phys. Rev. B **53**, 10532 (1996)
2.37 N. Yao, C. F. Klein, S. K. Behal, M. M. Disko, R. D. Sherwood, K. M. Creegan, D. M. Cox: Phys. Rev. B **45**, 11366 (1992)
2.38 W. I. F. David, R. M. Ibberson, T. J. S. Dennis, J. P. Hare, K. Prassides: Europhysics Lett. **18**, 219 (1992)
2.39 Z. C. Wu, D. A. Jelski, T. F. George: Chem. Phys. Lett. **137**, 291 (1987)
2.40 R. Heid, L. Pintschovius, J. M. Godard: Phys. Rev. B **56**, 5925 (1997)
2.41 L. T. Chadderton: J. Phys. Chem. Solids **54**, 1027 (1993)
2.42 D. P. Clougherty, J. P. Gorman: Chem. Phys. Lett. **251**, 353 (1996)
2.43 R. E. Stanton, M. D. Newton: J. Phys. Chem. **92**, 2141 (1988)
2.44 Z. Slanina, J. M. Rudzinski, M. Togasi, E. Osawa: J. Mol. Struct. **202**, 169 (1989)
2.45 F. Negri, G. Orlandi, F. Zerbetto: Chem. Phys. Lett. **144**, 31 (1988)
2.46 S. Baroni, P. Giannozzi, A. Testa: Phys. Rev. Lett. **58**, 1861 (1987)
2.47 P. Giannozzi, S. Baroni: J. Chem. Phys. **100**, 8537 (1994)
2.48 R. Car, M. Parrinello: Phys. Rev. Lett. **55**, 2471 (1985)
2.49 Q. M. Zhang, Jae-Yei Yi, J. Bernholc: Phys. Rev. Lett. **66**, 2633 (1991)
2.50 G. B. Adams, J. B. Page, O. F. Sankey, K. Sinha, J. Menéndez, D. R. Huffman: Phys. Rev. B **44**, 4052 (1991)
2.51 A. A. Quong, M. R. Pederson, J. L. Feldman: Solid State Commun. **87**, 535 (1993)
2.52 X. Q. Wang, C. Z. Wang, K. M. Ho: Phys. Rev. B **48**, 1884 (1993)
2.53 G. B. Adams, J. B. Page, O. F. Sankey, M. O'Keeffe: Phys. Rev. B **50**, 17471 (1994); G. B. Adams, J. B. Page, O. F. Sankey, M. O'Keeffe: Chem. Phys. Lett. **228**, 485 (1994); G. B. Adams, J. B. Page: in [2.6]

2.54 O. F. Sankey, D. Niklewski: Phys. Rev. B **40**, 3979 (1989); O. F. Sankey, D. J. Niklewski, D. A. Drabold, J. D. Dow: Phy. Rev. B **41**, 12750 (1990)
2.55 G. B. Adams, O. F. Sankey, J. B. Page, D. Drabold: Science **256**, 1792 (1992)
2.56 J. Harris: Phys. Rev. B **31**, 1770 (1985)
2.57 O. F. Sankey, R. E. Allen: Phys. Rev. B **33**, 7164 (1986)
2.58 B. P. Feuston: Phys. Rev. B **44**, 4056 (1991)
2.59 R. A. Jishi, R. M. Mirie, M. S. Dresselhaus: Phys. Rev. B **45**, 13685 (1992)
2.60 J. L. Feldman, J. Q. Broughton, L. L. Boyer, D. E. Reich, M. D. Kluge: Phys. Rev. B **46**, 12731 (1992)
2.61 W. Weber: Phys. Rev. B **15**, 1789 (1977)
2.62 G. Onida, G. Benedek: Europhys. Lett. **18**, 403 (1992)
2.63 S. Sanguinetti, G. Benedek, M. Righetti, G. Onida: Phys. Rev. B **50**, 6743 (1994)
2.64 K.-P. Bohnen, R. Heid, K. M. Ho, C. T. Chan: Phys. Rev. B **51**,5805 (1995)
2.65 W. Que, M. B. Walker: Phys. Rev. B **48**, 13104 (1993)
2.66 J. Yu, R. K. Kalia, P. Vashishta: Appl. Phys. Lett. **63**, 3152 (1993)
2.67 J. M. Pacheco, J. P. P. Ramalho: Phys. Rev. Lett. **79**, 3873 (1997)
2.68 Z.-H. Dong, P. Zhou, J. M. Holden, P. C. Eklund, M. S. Dresselhaus, G. Dresselhaus: Phys. Rev. B **48**, 2862 (1993)
2.69 K.-A. Wang, A. M. Rao, P. C. Eklund, M. S. Dresselhaus, G. Dresselhaus: Phys. Rev. B **48**, 11375 (1993)
2.70 M. C. Martin, X. Du, J. Kwon, L. Mihaly: Phys. Rev. B **50**, 173 (1994)
2.71 J. Menéndez, S. Guha: In Proc. 22nd Int. Conf. on the Physics of Semiconductors, ed. by D. Lockwood (World Scientific, Vancouver 1994) p. 2093
2.72 V. Schettino, P. R. Salvi, R. Bini, G. Cardini: J. Chem. Phys. **101**, 11079 (1994)
2.73 P. Bowmar, W. Hayes, M. Kurmoo, P. A. Pattenden, M. A. Green, P. Day, K. Kikuchi: J. Phys.: Condens. Matter **6**, 3161 (1994)
2.74 L. Akselrod, H. J. Byrne, S. Donovan, S. Roth: Chem. Phys. **192**, 307 (1995)
2.75 W. Brockner, F. Menzel: J. Mol. Struct. **378**, 147 (1996)
2.76 A. Graja, A. Lapinski, S. Król: J. Mol. Struct. **404**, 147 (1997)
2.77 C. Coulombeau, H. Jobic, P. Bernier, C. Fabre, D. Schütz, A. Rassat: J. Phys. Chem. **96**, 22 (1992)
2.78 J. R. D. Copley, D. A. Neumann, W. A. Kamitakahara: Cdn. J. Phys. **73**, 763 (1995)
2.79 C. Coulombeau, H. Jobic, C. J. Carlile, S. M. Bennington, C. Fabre, A. Rassat: Fullerene Sci. Technol. **2**, 247 (1994)
2.80 A. Sassara, G. Zerza, M. Chergui: J. Phys. B **29**, 4997 (1996)
2.81 W. Kraetschmer, K. Fostiropoulos, D. R. Huffman: Chem. Phys. Lett. **170**, 167 (1990)
2.82 D. S. Bethune, G. Meijer, W. C. Tang, H. J. Rosen, W. G. Golden, H. Seki, C. A. Brown, M. S. de Vries: Chem. Phys. Lett. **179**, 181 (1991)
2.83 R. Liu, M. V. Klein: Phys. Rev. B **45**, 11437 (1992)
2.84 B. Chase, N. Herron, E. Holler: J. Phys. Chem. **96**, 4262 (1992)
2.85 K. Sinha, J. Menéndez, G. B. Adams, J. B. Page, O. F. Sankey, L. D. Lamb, D. R. Huffman: In Appl. Spectr. Mater. Sci. **1437**, ed. by D. D. Saperstein (SPIE, Los Angeles 1991) p. 32
2.86 P. H. M. van Loosdrecht, P. J. M. van Bentum, M. A. Verheijen, G. Meijer: Chem. Phys. Lett. **198**, 587 (1992)

2.87 S. P. Love, D. McBranch, M. Salkola, N. V. Coppa, J. M. Robinson, B. I. Swanson, A. R. Bishop: Chem. Phys. Lett. **225**, 170 (1994)
2.88 L. Yulong, J. Yijian, L. Jingqing, M. Yujun, X. Sishen, Z. Zebo, Q. Shengfa: Phys. Rev. B **49**, 5058 (1994)
2.89 A. Rosenberg, C. Kendziora: Phys. Rev. B **51**, 9321 (1995)
2.90 J. Menéndez, J. B. Page: Bull. Am. Phys. Soc. **43**, 157 (1998)
2.91 D. W. Snoke, M. Cardona: Solid State Commun. **87**, 121 (1993)
2.92 M. Matus, H. Kuzmany: Appl. Phys. A **56**, 241 (1993)
2.93 P. H. M. van Loosdrecht, P. J. M. van Bentum, J. Meijer: Phys. Rev. Lett. **68**, 1176 (1992)
2.94 C. C. Homes, P. J. Horoyski, M. L. W. Thewalt, B. P. Clayman: Phys. Rev. B **49**, 7052 (1994)
2.95 C. C. Homes, P. J. Horoyski, M. L. W. Thewalt, B. P. Clayman, T. R. Anthony: Phys. Rev. B **52**, 16892 (1995)
2.96 P. J. Horoyski: Phys. Rev. B **54**, 920 (1996)
2.97 L. Pintschovius, B. Renker, F. Gompf, R. Heid, S. L. Chaplot, M. Haluska, H. Kuzmany: Phys. Rev. Lett. **69**, 2662 (1992)
2.98 S. A. Fitzgerald, A. J. Sievers: Phys. Rev. Lett. **70**, 3175 (1993)
2.99 P. J. Horoyski, M. L. W. Thewalt: Phys. Rev. B **48**, 11446 (1993)
2.100 P. J. Horoyski, M. L. W. Thewalt, T. R. Anthony: Phys. Rev. B **52**, R6951 (1995)
2.101 R. L. Cappelletti, J. R. D. Copley, W. A. Kamitakahara, F. Li, J. S. Lannin, D. Ramage: Phys. Rev. Lett. **66**, 3261 (1991)
2.102 K. Prassides, T. J. S. Dennis, J. P. Hare, J. Tomkinson, H. W. Kroto, R. Taylor, D. R. M. Walton: Chem. Phys. Lett. **187**, 455 (1991)
2.103 A. Sassara, G. Zerza, M. Chergui: Chem. Phys. Lett. **261**, 213 (1996)
2.104 I. Laszlo, L. Udvardi: Chem. Phys. Lett. **136**, 418 (1989)
2.105 F. Negri, G. Orlandi, F. Zerbetto: J. Chem. Phys. **97** 6496 (1992)
2.106 F. Negri, G. Orlandi, F. Zerbetto: J. Phys. Chem. **100**, 10849 (1996)
2.107 F. Negri: Private communication
2.108 A. Sassara, G. Zerza, M. Chergui, F. Negri, G. Orlandi: J. Chem. Phys. **107**, 8731 (1997)
2.109 G. Gensterblum, L.-M. Yu, J.-J. Pireaux, R. Caudano, J.-M. Themlin, S. Bouzidi, F. Coletti: Appl. Phys. A **56**, 175 (1993)
2.110 G. Gensterblum, J. J. Pireaux, P. A. Thiry, R. Caudano, J. P. Vigneron, P. Lambin, A. A. Lucas, W. Krätschmer: Phys. Rev. Lett. **67**, 2171 (1991)
2.111 S. H. Gallagher, R. S. Armstrong, P. A. Lay, C. A. Reed: Chem. Phys. Lett. **248**, 353 (1996)
2.112 S. H. Gallagher, R. S. Armstrong, W. A. Clucas, P. A. Lay, C. A. Reed: J. Phys. Chem. A **101**, 2960 (1997)
2.113 G. F. Bertsch, A. Smith, K. Yabana: Phys. Rev. B **52**, 7876 (1995)
2.114 K. Esfarjani, Y. Hashi, J. Onoe, K. Takeuchi, Y. Kawazoe: Phys. Rev. B **57** 223 (1998)
2.115 K.-J. Fu, W. L. Karney, O. L. Chapman, S.-M. Huang, R. B. Kaner, F. Diederich, K. Holczer, R. L. Whetten: Phys. Rev. B **46** 1937 (1992)
2.116 M. C. Martin, D. Koller, L. Mihaly: Phys. Rev. B **47** 14607 (1993)
2.117 R. Winkler, T. Pichler, H. Kuzmany: Zeit. Phys. B **96**, 39 (1994)
2.118 J. Onoe, K. Takeuchi: Phys. Rev. B **54**, 6167 (1996), as reported in [2.114]
2.119 T. Hara, J. Onoe, K. Takeuchi: J. Phys. Chem. B **101**, 9532 (1997)

2.120 S. Go, H. Bilz, M. Cardona: Phys. Rev. Lett. **34**, 580 (1975)
2.121 H. Bilz, D. Strauch, R. K. Wehner: *Handbuch der Physik* Vol. XXV, Pt. 2d (Springer, Berlin, Heidelberg 1984) p. 253
2.122 A. von Czarnowski, K. H. Meiwes-Broer: Chem. Phys. Lett. **246**, 321 (1995)
2.123 G. B. Adams, J. B. Page: unpublished
2.124 B. G. Johnson, J. Florián: Chem. Phys. Lett. **247**, 120 (1995)
2.125 D. Porezag, M. R. Pederson: Phys. Rev. B **54**, 7830 (1996)
2.126 S. Guha, J. Menéndez, J. B. Page, G. B. Adams: Phys. Rev. B **53**, 13106 (1996)
2.127 M. Wolkenstein: Dokl. Akad. Nauk. SSSR **32**, 185 (1941)
2.128 T. J. Dennis, J. P. Hare, H. W. Kroto, R. Taylor, D. R. M. Walton: Spectrochim. Acta **47A**, 1289 (1991)
2.129 P. R. Birkett, H. W. Kroto, R. Taylor, D. R. M. Walton, R. I. Grose, P. J. Hendra, P. W. Fowler: Chem. Phys. Lett. **205**, 399 (1993)
2.130 K. Lynch, C. Tanke, F. Menzel, W. Brockner, P. Scharf, E. Stumpp: J. Phys. Chem. **99**, 7985 (1995)
2.131 S. Lebedkin, H. Rietschel, G. B. Adams, J. B. Page, W. E. Hull, F. H. Hennrich, H.-J. Eisler, M. M. Kappas, W. Krätschmer: J. Chem. Phys. **110**, 11768 (1999)
2.132 D. W. Snoke, M. Cardona, S. Sanguinetti, G. Benedek: Phys. Rev. B **53**, 12641 (1996)
2.133 J. Lorentzen, S. Guha, J. Menéndez, P. Giannozzi, S. Baroni: Chem. Phys. Lett. **270**, 129 (1997)
2.134 K. T. Schomacker, J. K. Delaney, P. M. Champion: J. Appl. Phys. **85**, 4241 (1986)
2.135 W. Windl: *Ab-initio-Berechnung von Raman-Spektren in Halbleitern*, Dissertation, University of Regensburg, Germany (1995)
2.136 K. Sinha, J. Menéndez, R. C. Hanson, G. B. Adams, J. B. Page, O. F. Sankey, L. D. Lamb, D. R. Huffman: Chem. Phys. Lett. **186**, 287 (1991)
2.137 M. Matus, S. Balgavy, H. Kuzmany, W. Krätschmer: Resonance Raman Spectroscopy of Buckminsterfullerene, Physica C **185–189**, 423 (1991)
2.138 S. Guha, J. Lorentzen, K. Sinha, J. Menéndez, G. B. Adams, J. B. Page, O. F. Sankey: Mol. Cryst. Liq. Cryst. **256**, 391 (1994)
2.139 V. N. Denisov, B. N. Mavrin, G. Ruani, R. Zamboni, C. Taliani: Sov. Phys. JETP **75**, 158 (1992)
2.140 S. H. Gallagher, R. S. Armstrong, P. A. Lay, C. A. Reed: J. Am. Chem. Soc. **116**, 12091 (1994)
2.141 K. L. Akers, L. M. Cousins, M. Moskovits: Chem. Phys. Lett. **190**, 614 (1992)
2.142 A. Rosenberg: Phys. Rev. B **51**, 1961 (1995)
2.143 X. Andong, P. Haibin, Z. Xinyi, H. Yilin, F. Shaojun, S. Junyan, Z. Jian, X. Cunyi: J. Lumin. **63**, 301 (1995)
2.144 S. Leach, M. Vervloet, A. Desprès, E. Brèheret, J. P. Hare, T. J. Dennis, H. W. Kroto, R. Taylor, D. R. M. Walton: Chem. Phys. **160**, 451 (1992)
2.145 S. Larsson, A. Volosov, A. Rosén: Chem. Phys. Lett. **137**, 501 (1987)
2.146 W. Y. Ching, M.-Z. Huang, Y.-N. Xu, W. G. Harter, F. T. Chan: Phys. Rev. Lett. **67**, 2045 (1991)
2.147 Y. Wang, J. M. Holden, A. M. Rao, W.-T. Lee, X. X. Bi, S. L. Ren, G. W. Lehman, G. T. Hager, P. C. Eklund: Phys. Rev. B **45**, 14396 (1992)

2.148 P. Milani, M. Manfredini, G. Guizzetti, F. Marabelli, M. Patrini: Solid State Commun. **90**, 639 (1994)
2.149 A. C. Albrecht: J. Chem. Phys. **34**, 1476 (1961)
2.150 N. Laouini, O. K. Andersen, O. Gunnarson: Phys. Rev. B **51**, 17446 (1995)

3 Raman Scattering from Surface Phonons

Norbert Esser and Wolfgang Richter

With 38 Figures

During recent decades Raman spectroscopic equipment has undergone an enormous improvement, mainly due to the advent of reliable and tunable cw laser light sources and sensitive, low noise multichannel light detection systems. Naturally, the technical advances have stimulated the application of Raman spectroscopy and expanded its capabilities. Raman scattering has become a widespread technique used in many areas of fundamental research [3.1, 3.2] and in the field of materials characterization [3.3, 3.4], in particular in an industrial context. This strong interest in Raman spectroscopy originates from the fact that a broad spectrum of information about solid state excitations (phonons, magnons, plasmons, single electronic excitations) is accessible and can be exploited for the solution of fundamental as well as more applied questions [3.2, 3.3, 3.5, 3.6, 3.7].

While Raman spectroscopy is a standard technique for investigating lattice dynamical properties of bulk semiconductors, it has been only recently demonstrated that the sensitivity of modern Raman equipment even allows the observation of Raman signals generated by surface (or interface) phonons. With this development, a new field of application, namely the characterization of surface vibrational properties, has opened for Raman spectroscopy.

Experimental techniques for analyzing surface phonon properties have so far been based on surface sensitive probes such as low-energy electrons and atoms, i.e., High-Resolution Electron Energy-Loss Spectroscopy, (HREELS) and Helium Atom Scattering (HAS). Due to the strong interaction of low-energy electrons and atoms with matter, the penetration depth of these probes is limited to the few outermost atomic layers giving rise to the surface sensitivity of these techniques. Raman spectroscopy, on the other hand, is based on the comparably weak interaction of photons with matter. Even under strong absorbing conditions the penetration depth of photons is at least 10 nm (approximately fifty atomic layers) and thus the Raman signal is bulk rather than surface related. Nevertheless, since usually surface and bulk vibrational modes should differ in their respective eigenfrequencies, the surface related information might be extracted utilizing the spectral selectivity of Raman spectroscopy. The discrimination of bulk and surface contributions, of course, will be most easily performed in cases where the surface eigenfrequencies are significantly different from the bulk values and do not overlap with the bulk Raman spectrum. This has favored the study of surfaces terminated

Topics in Applied Physics, Vol. 76
Light Scattering in Solids VIII
Eds.: M. Cardona, G. Güntherodt

by different kinds of atoms to those present in the bulk (e.g., Sb/GaAs(110), As/Si(111), H/Si(111)). Most experimental results discussed in this article deal with such hetero-terminated surfaces. Nevertheless, the observation of surface phonons on non-hetero-terminated (clean) surfaces is possible, a fact that was recently demonstrated for the first time [3.8]. On clean InP(110) surfaces interesting new results were obtained utilizing the high-resolution potential of Raman spectroscopy to study surface phonons.

In general, due to the large penetration depth of light, the Raman signal from surface vibrations should be small as compared to bulk contributions. However, as well known from the study of bulk properties, the Raman scattering cross section undergoes strong variations, of several orders of magnitude, depending on the possible resonance of the probing light with excitations in the electronic band structure of the solid. Since the electronic structure of the surface in general is distinct from the bulk electronic band structure, under favorable conditions (surface state resonance) the Raman signal from *surface* vibrations may increase considerably with respect to that of the volume. As we will show, this might further help to discriminate the surface contribution against that of the bulk.

In the following we will summarize the present understanding of Raman scattering from phonons at semiconductor surfaces and the experimental results available so far. The surfaces are assumed to possess two-dimensional long range order. Similar considerations will be assumed for heterointerfaces, where one just deals with solid–solid instead of solid–vacuum interfaces. However, experimental as well as theoretical interest has not focussed much on localized interface phonons up to now; only a smaller part of this article (Sect. 3.6) will be devoted to them. We would like to emphasize here that this article will not deal with the vibrational properties of molecules adsorbed on surfaces. This is actually a large field where Raman spectroscopy has been applied either as normal Raman scattering [3.9] or through the by now famous Surface Enhanced Raman Scattering (SERS) (SERS) [3.10, 3.11, 3.12]. Nearly all of the SERS work deals with molecules adsorbed on metals, with the emphasis on catalysis or in understanding processes on electrodes. The main focus in these papers is therefore on molecular vibrational properties with less emphasis on the substrate or a possible long range order. The most important results in this field have been reviewed recently by *Suetaka* [3.13].

To define the subject of the present chapter we will briefly discuss the lattice dynamics of surfaces in Sect. 3.1. Thereafter, in Sect. 3.2, the well-known basic interaction mechanisms of Raman scattering in solids, which have been described in detail in a number of review articles [3.2, 3.3, 3.6, 3.7, 3.14, 3.15], are summarized and extended to the surface case. After this introductory sections examples of Raman scattering from surface phonons are discussed. Experimental results are compared with calculations of eigenfrequencies and Raman cross sections of surface modes. The Sb/III–V(110) monolayer terminated surfaces will serve as a model structure to establish a microscopic understanding of the basic interaction mechanisms. Other examples in the following sections are discussed on the basis of this prototype system.

3.1 Surface Phonons

Surface phonons represent the vibrational modes of well ordered surfaces with translational symmetry. As a consequence, the wave vector $\boldsymbol{k}_{||}$ (in the surface plane) is a well defined quantum number and the vibrational spectrum is described within the Surface Brillouin Zone (SBZ). To distinguish three- and two-dimensional k-space, the high symmetry points of the SBZ are written with an overline (e.g., $\bar{\Gamma}$ instead of Γ).

The detection and analysis of surface phonons by Raman scattering is still a very new and exciting field of research. In this section we will briefly sketch the present status of the theory of surface phonons and summarize the specific characteristics of the standard experimental methods for surface vibrational analysis. More detailed reviews can be found in [3.17,3.18,3.19,3.20,3.21].

3.1.1 Dispersion of Surface Phonons

In general, solid surfaces are expected to exhibit a vibronic spectrum different from the bulk phonon dispersion relations, due to lower dimensionality, different atomic structure and chemical bonding. These differences in the vibrational properties should sensitively depend on the detailed atomic structure of the first few layers at the surface. Their analysis by appropriate theoretical methods can therefore help in understanding surface dynamics and clarify structural properties of surfaces.

The first theoretical approach to surface vibrational properties dates back to the last century with the theory of elastic surface waves given by *Lord Rayleigh* in connection with the propagation of earthquake waves [3.22]. This macroscopic ansatz soon developed and expanded into theoretical work describing surface waves in semi-infinite continua and was successfully applied to the description of surface acoustic waves. The first microscopic calculation on a level equivalent to the Born-von Karman lattice dynamical theory [3.23] was performed much later by *Lifshitz* and *Rosenzweig* in 1948 [3.24] by using the Green's function method to describe the free surface as a perturbation of the bulk. This approach, as well as an alternative method using trial functions localized at the surface [3.18,3.21], was widely used for over two decades. However, with the development of increased computational power over the last twenty years, the direct numerical diagonalization of the dynamical matrix of an appropriate slab system has become popular [3.18,3.21,3.28]. Thereby, the semiinfinite crystal is replaced by an infinite number of slabs which obey translational symmetry along the surface normal. Each slab consists of a set of infinitely extended two-dimensional layers, with vacuum between them and a sufficient thickness to yield independence of the surface localized phonon

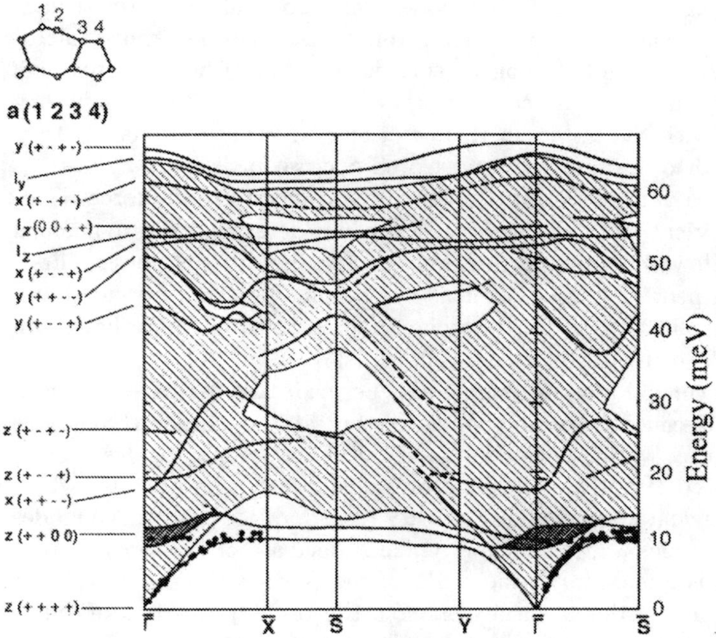

Fig. 3.1. Surface phonon dispersion curves for Si(111) (2 × 1). The filled circles represent experimental data from HAS, lines are calculations based on a 24 layers slab. The solid lines represent surface localized modes, broken lines are surface resonances, the hatched areas are surface projected bulk states. The labels at the left side denote the atomic displacements of the surface eigenmodes at $\bar{\Gamma}$ for the numbered top layer atoms. The acoustic surface mode $z(++++)$ is the so-called Rayleigh mode [3.17]

modes from the number of atomic layers (usually between 10 and 30). With these model structures three dimensional periodicity is restored but with unit cells containing a much larger number of atoms (= atoms per surface unit cell × number of atomic layers in the slab). This technique was essentially developed by *de Wette* [3.17] and represented a pioneering step towards realistic calculations of surface phonon dispersion curves. These calculations, when performed with an increasing number of layers per slab, show in a very instructive manner the development of bulk and surface modes; as one adds more and more layers the increasing number of modes form either well separated single modes or dense bands. The latter correspond to the bulk modes projected onto the SBZ whereas the former represent surface modes.

A typical result of such calculations is displayed in Fig. 3.1 together with HAS data. Surface modes are found whose penetration depth in the continuum limit (small wave vector) is quite large, e.g., in the order of their in-plane wavelength. They are termed macroscopic surface modes. Examples are the already mentioned Rayleigh modes and the Fuchs-Kliewer surface polariton modes [3.18,3.21,3.25]. Other modes which are confined to the surface within a distance equivalent to few interatomic distances are termed microscopic [3.26, 3.27]. In the following we specifically will deal with microscopic surface modes. If microscopic modes overlap in energy with bulk bands they may hybridize with the latter. In this case, termed surface resonance, they are less localized to the surface than isolated microscopic surface modes.

A crucial task is the description of the interatomic restoring forces. Traditionally, empirical methods have been used to determine the dynamical matrix [3.17]. However, through the increase of computational power a first-principles treatment of surface phonons has become possible. Recent examples are the application of the interplanar force constant and the frozen phonon method to III–V(110) surfaces [3.29, 3.30].

Some more general statements can also be made about surface phonons at this stage. Because there are bonds missing at the surface the surface phonon frequencies will be in general lower than those of bulk modes. This is globally expressed by the first moment of the energy density of states which is expected to give a lower center of gravity (centroid) for the surface modes than the bulk modes and also shows up in a decrease of the second order moment which is a measure of the width of the density of states [3.19]. For the experimental detection of surface phonons the former property is of course not advantageous since it makes surface phonons overlap with the bulk and therefore more difficult to detect. The smaller width of the density of states, giving sharper structures and higher peak values, may be a positive feature with respect to detectability. As a consequence of the smaller frequencies the amplitudes of the atomic displacements will be larger for surface than for bulk phonons and thus increase the detectability. This effect probably should be more significant for displacements perpendicular to the surface (in a direction where bonds are missing) than for parallel displacements.

3.1.2 Experimental Methods

Experimental results concerning surface phonons have for a long time lagged behind their theoretical description. For bulk phonon dispersion measurements *Neutron scattering* is the standard method [3.31]. However, because of the large penetration depth of neutrons, and their low scattering cross section, it is not appropriate for surface phonon measurements. Nevertheless a few reports on systems with large surface areas have appeared [3.32, 3.33]. In contrast, nearly all studies on surface phonons have been performed by HREELS and also by HAS.

In HREELS, monochromatic low energy electrons scattered from a surface under UHV conditions are analyzed with respect to their energy distribution and their scattering angle. The most prominent scattering mechanism for low energy electrons (kinetic energy below 10 eV) is dipole scattering mediated by the time dependent electric field of the propagating electrons. By this mechanism, often applied to study adsorbate vibrations, dipole active surface vibrational modes at $\boldsymbol{k}_{||} \approx 0$ are detected with high sensitivity. $\boldsymbol{k}$-resolved studies of surface phonons become possible through utilization of the comparably much weaker impact scattering mechanism between electrons and the ion cores in the solid which allows one to realize large $\boldsymbol{k}_{||}$ values in the order of a reciprocal lattice vector at kinetic energies of 100–300 eV. The surface sensitivity in the impact scattering regime is large, since the inelastic mean free path of the electrons amounts to only few atomic layers. A considerable drawback of the HREELS technique, however, is the low energy resolution. Due to the strong interaction of low-energy electrons with stray magnetic and electric fields the realization of high-resolution analyzers with sufficient transmission is still a challenge. Nevertheless, recently developed spectrometers achieve, under impact scattering conditions, an energy resolution around 2–3 meV on metal surfaces [3.36].

It should be mentioned, however, that the energy resolution may strongly depend on the structural and electrostatic homogeneity of the surface [3.21, 3.34, 3.35], a fact which makes especially the study of compounds and nonmetals difficult.

HAS, on the other hand, overcomes the drawback of resolution and offers an extreme surface sensitivity. The low-energy helium atoms (8–60 meV) used for HAS studies are in fact reflected by the valence charge distribution at the outermost surface layer. The energy resolution provided by time-of-flight analyzers is excellent, in the range of 0.2 to 1 meV. Due to the short range interaction between He atoms and the surface potential, all vibrational modes which give rise to a distinct modulation of the surface charge distribution are observable, and large $\boldsymbol{k}_{||}$ momentum transfers, similar to those in HREELS, become possible. The main drawback of HAS, besides the rather complex experimental setup, is its limitation to low energy transfers (up to 30 meV) and, consequently, low vibrational eigenfrequencies. Thus, excellent data on acoustic surface phonon modes have been collected, but higher frequency optical modes are seldom reported [3.21, 3.37, 3.38].

A few investigations on surface phonons have been performed by Infrared Reflection Absorption Spectroscopy (IRAS) [3.13, 3.39]. In contrast to HREELS and HAS, IRAS is an optical technique which may operate under UHV conditions as well as in a gas phase environment. Only dipole active surface vibrations at the Brillouin zone center are detectable. IRAS has been applied mainly in the study of dipole allowed vibrations of adsorbed molecules. While IRAS provides a high energy resolution of better than 0.05 meV this technique suffers from a comparably low surface sensitivity. Studies of adsorbate vibrations are normally performed using either modulation techniques or complex arrangements with multiple internal reflections to enhance the surface sensitivity [3.13, 3.39].

Thus, for the study of surface phonons and their dispersion, the two standard techniques have been HAS and HREELS. The two techniques are complementary: HAS provides superior energy resolution but is limited to phonons of energies below 30 meV, whereas HREELS allows one to study also higher energy phonons, e.g., optical modes, but with comparably low resolution [3.18, 3.21].

The study of microscopic optical surface modes with high spectral resolution still remains a challenge. For bulk phonon modes Raman scattering has provided its high resolution capability to complement neutron scattering measurements. It should also be able to fulfill this role for surface phonons. Raman scattering is sensitive to higher frequency optical phonon modes. The momentum transfer is somewhat larger than in IRAS (visible light has larger wave vectors than infrared radiation), but is still very close to the Brillouin zone center. As an optical technique operating with light in the visible range it is easily adaptable to UHV as well as to gas phase surroundings and offers a high spectral resolution of better than 0.01 meV. Since the scattering process connects two photons (photon-in, photon-out) it is one order of perturbation theory higher in the electric fields than IRAS (just photon-in). The interaction with matter leading to the scattering of light is therefore relatively weak and, for that reason, Raman scattering had not been used until recently for the analysis of surface phonons. However, in contrast to IRAS where photons couple directly to the lattice at the vibrational eigenfrequencies, in Raman scattering the emitted and adsorbed photons interact with electronic excitations whose energy is modulated by the lattice vibration. The photon-electron interaction may lead to a large enhancement of the scattering cross section due to resonances of the photons with electronic interband transitions. Under favorable resonant conditions the Raman signal may thus become enhanced by several orders of magnitude without the need of any special sample arrangement. This effect, called Resonance Raman Scattering, has been known for many years in bulk semiconductors [3.2, 3.14]. In the following sections several examples will demonstrate that the resonance effect can be exploited for the study of phonon modes of semiconductor surfaces. The resonances will then concern electronic surface states and, like in the bulk case, may lead to surprisingly high scattering cross sections. As a consequence, Raman scattering can give access to zone center optical surface modes with unparalleled high spectral resolution. Like in the case of bulk phonons, many specific questions concerning surface phonons may then be addressed by Raman scattering, e.g., the dependence of eigenfrequencies and halfwidths on real surface structure (degree of order or kind and number of impurities) or the influence of external perturbations (such as stress, temperature) [3.18, 3.19, 3.21].

The observation of vibrational modes by Raman spectroscopy requires, of course, symmetry selection rules to be fulfilled (i.e., Raman activity must be present). However, since the two-dimensional point groups of surfaces contain less symmetry elements than the ones of the bulk, the Raman selection rules impose no serious limitations for the observability of surface phonons. Nevertheless, the selection rules are sufficiently significant to derive information about the symmetry of the vibrational modes. In contrast to HREELS and HAS, Raman scattering is not intrinsically a surface sensitive probe, since the photons penetrate relatively deep into the bulk (penetration depth larger than 10 nm). Thus, surface related information must be discriminated against the stronger bulk contributions, based on the fact that surface electronic and vibronic properties are distinct from those of the bulk. By tuning into electronic surface resonances the surface signal may be selectively enhanced and may be distinguished from bulk contributions, especially if the respective phonon frequencies are spectrally separated. This separation is naturally promoted on surfaces which are terminated by different kinds of atoms, not part of the bulk material (e.g., H, As, Sb). Comparison to the clean surface and controlled change of surface termination allows for identification of the surface phonons. In the study of clean surfaces, vice versa, a controlled modification of the surface, for instance by oxidation, may serve to separate surface from bulk contributions. Especially for the case of microscopic surface phonon modes the dynamical properties of the outermost few atomic layers can easily be probed by this differential procedure, in spite of the much larger photon penetration depth.

In Sect. 3.2 the basics of Raman scattering, the resonance effects on the Raman cross section, and the origin of the symmetry selection rules are briefly summarized and discussed with respect to Raman scattering by surface phonons. The experimental realization of a setup for surface Raman studies is shown. In Sect. 3.3 Raman scattering from surface phonons is treated in detail for the well-characterized model system of III–V(110) surfaces terminated by a Sb-monolayer. Further examples of surface Raman scattering for As/Si(111), H/Si(111) and S/InP(110) are summarized in Sect. 3.4. Surface phonons on clean (110) surfaces of III–V-semiconductors are treated in Sect. 3.5 and finally, results obtained for microscopic interface modes are discussed in Sect. 3.6.

3.2 Fundamentals of Raman Spectroscopy

The basic features of Raman scattering from surface phonons and from bulk phonons are very similar. This concerns the experimental techniques, the kind of selection rules, resonance phenomena and other properties. For this reason we will confine this section mainly to the differences occurring in Raman scattering between surface and bulk phonons. For the more general aspects of Raman scattering in solids the reader is referred to the many reviews available [3.1, 3.2, 3.14].

The main difference between Raman scattering by surface phonons and bulk scattering is that one is dealing with a surface – a 2-dimensional system – which has different electronic states and phononic properties than the bulk. While these differences do not cause any principal limitations, the small number of atoms participating in the surface scattering process, as compared to bulk scattering, turns out to be the main experimental obstacle. As a result the scattering experiment should be carefully designed. Besides standard optical measures such as high aperture, optimized collecting optics for the scattered light between the UHV chamber and the monochromator, it turns out that the main advantage comes from the exploitation of cross-sectional resonance enhancements. For this reason previous knowledge about the electronic surface band structure is extremely helpful and greatly simplifies the experiment. Therefore, in the following section dealing with Raman scattering from specific surfaces, discussion of the electronic band structure of the surfaces under consideration plays a very important role.

3.2.1 Energy and Wave-Vector Conservation

Raman spectroscopy is understood as the spectral investigation of inelastically scattered light ($\hbar\omega = 1$–$6\,\mathrm{eV}$) with energy transfers larger than approximately $1\,\mathrm{meV}(\ 8\,\mathrm{cm}^{-1})$. Grating monochromators (double or triple) with high resolution and more importantly, with high contrast, are usually employed for the spectral analysis. In the scattering process a certain amount of energy is gained or lost by an incident photon with energy $\hbar\omega_\mathrm{i}$ (incident) in order to create or annihilate elementary excitations of the solid (here surface phonons), resulting in a scattered photon of a different energy $\hbar\omega_\mathrm{s}$ (scattered). The amount of energy transferred corresponds to the eigenenergy $\hbar\omega_j$ of the elementary excitation involved labelled by "j".

$$\hbar\omega_\mathrm{s} = \hbar\omega_\mathrm{i} \pm \hbar\omega_j . \tag{3.1}$$

Here the "minus" sign stands for a phonon excitation (Stokes process) while the "plus" sign implies a phonon annihilation (anti-Stokes process). Because of their higher intensity resulting from the relation:

$$I_\mathrm{Stokes}/I_\mathrm{anti-Stokes} = \exp(\hbar\omega_j/kT)\ , \tag{3.2}$$

in most cases only Stokes processes are studied. This fact also holds for surface phonons.

Pseudo-momentum ($\hbar\boldsymbol{k}$) conservation determines the phonon wave vector $\boldsymbol{k}_j$ involved. The wave vector $\boldsymbol{k}_j$ transferred to the surface phonon excitation is related to the wave vector of the incident $\boldsymbol{k}_\mathrm{i}$ and scattered light $\boldsymbol{k}_\mathrm{s}$ according to:

$$\boldsymbol{k}_\mathrm{s} = \boldsymbol{k}_\mathrm{i} \pm \boldsymbol{k}_j , \tag{3.3}$$

where only the components of the light wave vectors parallel to the surface have to be considered for surface phonons.

Both conservation laws select well-defined (ω, $\boldsymbol{k}_j$) pairs out of the whole range given by the dispersion relations. The specific phonon excitation observed in a scattering experiment depends on the parameters of the process, i.e., the photon frequencies and the directions of incident (laser beam) and scattered light (i.e., the position and aperture of the collection lenses). Because a small penetration depth of the light is desired in order to discriminate more easily against bulk phonon scattering, the experiments are always performed in spectral regions with strong absorption and consequently in near-backscattering geometry. Thus the components of the light wave vectors parallel to the surface are even smaller than those for bulk scattering, and are of the order of 1 % of the size of the SBZ. Thus only surface phonons very close to the center of the Brillouin Zone may be probed.

3.2.2 Scattering Intensity

Raman scattering, involving two electromagnetic fields with different frequencies, may be discussed in the form of a higher order dielectric susceptibility [3.2, 3.14, 3.15, 3.40]. This has been termed transition susceptibility [3.41], a name which refers to the fact that, in addition to the electromagnetic fields, the creation or annihilation of an elementary excitation is involved.

The standard linear dielectric susceptibility (tensor) $\tilde{\chi}(\omega, \boldsymbol{k})$ describes the first order interaction of an electromagnetic wave $\boldsymbol{E}(\omega, \boldsymbol{k})$ with matter. Due to the small $\boldsymbol{k}$ vector of visible light the explicit $\boldsymbol{k}$ dependence can be neglected in most cases, and the induced polarization reads:

$$\boldsymbol{P}(\omega) = \epsilon_0 \tilde{\chi}(\omega) \boldsymbol{E}(\omega). \tag{3.4}$$

Raman scattering processes can then be visualized in one of three ways:

(i) in a semiclassical manner, as induced by the periodic modulation of the dielectric susceptibility (polarizability theory),

(ii) in a quantum-mechanical picture of the susceptibility, where oscillator strengths or transition energies associated with electronic interband transitions are modified by the lattice deformation in a quasistatic manner, or

(iii) finally, in a perturbative approach including the electron–phonon interaction in addition to the electron–photon interactions of the incident and scattered fields [3.6].

The generation of scattered light with frequency ω_s by incident light with frequency ω_i may be expressed in either case using the transition dielectric susceptibility tensor $\tilde{\chi}(\omega_\mathrm{i}, \omega_\mathrm{s})$ that connects the exciting electromagnetic field with frequency ω_i to the scattered field with frequency ω_s:

$$\boldsymbol{P}(\omega_\mathrm{s}) = \epsilon_0 \tilde{\chi}(\omega_\mathrm{i}, \omega_\mathrm{s}) \boldsymbol{E}(\omega_\mathrm{i}). \tag{3.5}$$

Here $\boldsymbol{P}(\omega_\mathrm{s})$ is the oscillating polarization which gives rise to the scattered light wave and $\boldsymbol{E}(\omega_\mathrm{i})$ is the oscillating electric field of the incident light wave.

The scattered intensity can finally be expressed as dipole radiation using this generalized dielectric susceptibility $\tilde{\chi}(\omega_\mathrm{i}, \omega_\mathrm{s})$:

$$I_\mathrm{s} = I_\mathrm{i} \frac{\omega_\mathrm{s}^4 V}{(4\pi\epsilon\epsilon_0)^2 c_0^4} |\boldsymbol{e}_\mathrm{s} \epsilon_0 \tilde{\chi}(\omega_\mathrm{i}, \omega_\mathrm{s}) \boldsymbol{e}_\mathrm{i}|^2 \tag{3.6}$$

I_i, I_s and $\boldsymbol{e}_\mathrm{i}$, $\boldsymbol{e}_\mathrm{s}$ denote the intensity and polarization of incident and scattered light. From there the scattering efficiency, with the dimension of an inverse length, is defined as [3.2, 3.14]:

$$S_{\alpha,\beta} = \frac{\mathrm{d}I^\mathrm{s}_\alpha}{LI^\mathrm{i}_\beta \mathrm{d}\Omega} = \frac{|\tilde{\chi}_{\alpha,\beta}(\omega_\mathrm{i}, \omega_\mathrm{s})|^2 \omega_\mathrm{s}^4 V}{c_0^4} \tag{3.7}$$

where L is the length of the total scattering volume in the direction of the propagation of the incident light, Ω is the solid angle under which the scattered light is collected, and α, β denote the polarization of incident and scattered light. In bulk phonon scattering L is the penetration depth of the light (or the whole sample dimension if the sample is transparent). In surface phonon scattering, in contrast, L is given by the thickness of the "surface", e.g., the localization depth of the surface phonon which amounts to a few monolayers in general.

The generalized susceptibility tensor can be related to the linear susceptibility by a Taylor expansion in terms of the lattice deformation, assuming a modulation of the linear susceptibility proportional to generalized coordinates Q_j which correspond to the lattice deformation caused by the phonon excitation:

$$\begin{aligned} \chi_{\alpha,\beta}(\omega_\mathrm{i}, \omega_\mathrm{s}) = {} & \chi^0_{\alpha,\beta}(\omega_\mathrm{i}) \\ & + \sum_j Q_j \cdot \left(\frac{\partial \chi_{\alpha,\beta}(\omega_\mathrm{i})}{\partial Q_j} \right) \\ & + \sum_{j,j'} Q_j Q_{j'} \cdot \frac{1}{2} \left(\frac{\partial^2 \chi_{\alpha,\beta}(\omega_\mathrm{i})}{\partial Q_j \, \partial Q_{j'}} \right) \\ & + \ldots . \end{aligned} \tag{3.8}$$

Here Q_j is the phonon amplitude in terms of a generalized coordinate. The term with the first derivative in (3.8) describes one phonon scattering, that with second order derivatives two-phonon scattering, and so on [3.2, 3.14, 3.15].

In a microscopic quantum mechanical approach the light scattering process may be described using time-dependent perturbation theory [3.44]. The dominant term amounts to [3.6]:

$$\chi_{\alpha,\beta}(\omega_\mathrm{i}, \omega_\mathrm{s}) = \frac{e^2}{m_0^2 \cdot \omega_\mathrm{s}^2 \cdot V} \sum_{e,e'} \frac{\langle 0|p_\alpha|e'\rangle \langle e'|H_\mathrm{E-L}|e\rangle \langle e|p_\beta|0\rangle}{(E_{e'} - \hbar\omega_\mathrm{s})(E_e - \hbar\omega_\mathrm{i})} \tag{3.9}$$

where m_0 is the electron mass, V the scattering volume, p_α, p_β the cartesian components of the momentum operators, E_e, $E_{e'}$ the energies of the excited electron–hole pair states and $H_\mathrm{E-L}$ the electron–phonon interaction Hamiltonian.

Equation (3.9) includes the transition from the ground state $|0\rangle$ to an excited electronic state $|e\rangle$ (photon absorption), scattering of the generated electron–hole pair into another state $|e'\rangle$ via electron–lattice interaction, and finally the transition back to the electronic ground state $|0\rangle$ under photon emission. It should be noted that the $1/\omega_s^2$ pre-factor appearing in (3.9) is due to the $\boldsymbol{p}\cdot\boldsymbol{A}$ coupling used in the electron–photon interaction. As a consequence the ω_s^4 dependence originating from the dipole radiation in (3.7), which describes the frequency dependence of the Raman cross section observed for exciting frequencies below all the electronic transitions energies, does not appear explicitly. It has been shown, however, that the ω_s^4 dependence is analytically regained by using the $\boldsymbol{r}\cdot\boldsymbol{E}$ (dipole approximation) instead of the $\boldsymbol{p}\cdot\boldsymbol{A}$ formalism in calculating the electron–photon interaction [3.42].

For photon energies not too close to electronic resonances (i.e., further away than the linewidth of the electronic transition) the result of (3.9) has been shown within a quasistatic approximation ($\omega_j \to 0$) to be equivalent to the expression in (3.8) containing the derivative of the dielectric susceptibility if one uses the quantum mechanical description of the susceptibility [3.2,3.14].

If only two bands are involved in the scattering process, (3.9) can be written in the form:

$$\chi_{\alpha,\beta}(\omega_\mathrm{i},\omega_\mathrm{s}) = D_j \cdot \frac{\mathrm{d}\chi_{\alpha,\beta}(\omega)}{\mathrm{d}(\hbar\omega)} \tag{3.10}$$

where D_j is the corresponding electron–phonon interaction matrix element for the surface electronic state and the surface phonon under question. The derivation assumes also that this matrix element is constant for all electronic transitions involved. Only under such an assumption can it be taken in front of the summation of (3.9) so that the remaining sum then just represents the derivative of the susceptibility with respect to energy.

The relationships (3.9) and (3.10) can be utilized to calculate the Raman scattering cross section. The transition susceptibility $\tilde{\chi}(\omega_\mathrm{i},\omega_\mathrm{s})$ is given by the derivative of the linear susceptibility $\tilde{\chi}(\omega)$ with respect to energy (3.10) or, more generally, with respect to the lattice deformation (3.8). The Raman cross section can consequently be obtained via band structure calculations, by taking the difference of the susceptibilities for the equilibrium lattice and that obtained after a shift of atomic positions according to the phonon normal coordinates [3.2, 3.14, 3.45]. The energy derivative is also quite useful in situations where experimental data for the susceptibility are available. This is quite often the case since surface dielectric functions can be determined by ellipsometric measurements.

The electronic properties that may be influenced by a phonon are eigenenergies and eigenfunctions of the electronic states. If the mechanical lattice deformation caused by the phonon is the only microscopic origin of the modulation of electronic properties, the interaction mechanism is called Deformation Potential (DP) scattering. This has been intensively discussed for bulk scattering with photon energies at different energy gaps in III–V semiconductors [3.14].

An additional interaction mechanism is involved for the longitudinal vibration modes of IR-active bulk optical phonons, i.e., phonons for which the symmetry of their eigenvectors allows a dipole moment per unit cell [3.47,3.48]. The longitudinal component of such phonons generates, for small wave vectors, a macroscopic, long-range electric field in the three-dimensional case. Besides acting as an additional restoring force and thus leading to an increased frequency of the longitudinal optical phonon with respect to the transverse partner, this field may also give rise to an additional light scattering mechanism since it may interact directly with the electrons. This interaction mechanism is called *Fröhlich interaction* [3.49].

In two dimensions, however, the corresponding electric field goes to zero in the small wave vector limit as $\sqrt{(k_j)}$ (nonretarded limit) or as k_j (retarded limit) [3.43]. Thus any LO-TO splitting should disappear and consequently the Fröhlich interaction should be of little or no importance. Indeed up to now experiments have not indicated such a scattering contribution, which in the bulk case manifests itself by specific selection rules and strong resonances with states involving excitonic contributions.

3.2.3 Resonance Effects

Apart from the w_s^4 dependence corresponding to the dipole radiation, the Raman scattering cross section will show a pronounced dependence on the energy of the exciting photons. Maxima in the Raman cross section will occur for photon energies matching critical points of the electronic band structure (3.5,3.8,3.9). This condition, called Resonant Raman Scattering (RRS), (RRS), has been intensively exploited for bulk phonon scattering [3.2,3.14] and can be readily applied to the surface electronic band structure.

As already mentioned, the enhancement of Raman cross section under resonant conditions is of crucial importance for surface phonons in order to enable detection of Raman signals from the extremely thin layers (a few ML). Moreover, the determination of the spectral dependence of the Raman cross section can be used to analyze selectively the electronic properties of the surface where the scattered light is generated. This can be accomplished by varying the exciting photon energy and monitoring the Raman efficiency. In such a way the electronic surface band structure and the corresponding electron–phonon interactions may be studied in a similar way as for the bulk [3.2,3.14]. Examples will be discussed in Sect. 3.3.5.

In the susceptibility approach the Raman resonance is reflected in an enhancement of the derivative of the linear susceptibility (3.10) at critical points of the electronic band structure. The validity of the susceptibility model can thus be experimentally verified by comparing the experimental result to the derivative of the susceptibility. This, in fact, has been done for several bulk semiconductors such as GaAs [3.50], InP [3.51] and Si [3.52]. The susceptibility may be obtained also for surfaces at least in the case of hetero-terminated surfaces from linear optical experiments like ellipsometry or reflectance [3.15, 3.16] or may be calculated from the corresponding electronic band structure.

3.2.4 Selection Rules

Selection rules are the result of symmetry considerations for the polarizations of incident and scattered fields usually derived by group theory. They determine which components of the Raman tensor $\tilde{\chi}$ are nonzero for a certain phonon normal mode. Such an analysis has been performed for all irreducible representations of the 32 crystallographic point groups. The non-zero components of these tensors can be found, for example, in [3.2, 3.55]. For the two-dimensional lattices Raman tensors have not been listed, to our knowledge. However, since the symmetry groups for two-dimensional lattices form a subset of the corresponding three-dimensional ones, it is easy to make use of the three-dimensional Raman tensors. After having determined the symmetry point group of the surface unit cell (or, more precisely, the factor group of the space group) and the phonon normal coordinates (or, at least, the corresponding irreducible representations), one can take then the corresponding x–y subspace of the three-dimensional Raman tensors and label them as two-dimensional Raman tensors for the case of surface phonon Raman scattering.

Since the two-dimensional lattices contain fewer symmetry operations than the three-dimensional groups it follows that the symmetry groups, and the number of irreducible representations, are smaller. Therefore, Raman selections rules for surface phonons pose few restrictions and usually the vibrational modes are Raman active. Nevertheless, they allow, as we will show, the experimental determination of different phonon eigenmodes.

Experimentally, certain tensor components are selected by choosing the polarization directions $\boldsymbol{e}_{\mathrm{i}}$ and $\boldsymbol{e}_{\mathrm{s}}$ of incident and scattered light along appropriate crystal symmetry directions. The experimental configuration will be given in the PORTO notation as $\boldsymbol{k}_{\mathrm{i}}(\boldsymbol{e}_{\mathrm{i}};\boldsymbol{e}_{\mathrm{s}})\boldsymbol{k}_{\mathrm{s}}$ [3.2, 3.14].

Group theory may be useful not only for an analysis of the macroscopic symmetry properties leading to Raman tensors but also, and especially, for finding the dominant resonance contributions. A microscopic analysis of the Raman process with regard to the steps occurring in the third order perturbation expression of (3.9) may provide this information. The irreducible representations of the relevant electronic surface states have thus to be known from surface band structure calculations. Determining then the irreducible representations in the product representations of the phonon and the different possible electronic states, according to the matrix elements in (3.9), quite often allows one to pinpoint the electronic transition responsible for the Raman resonance [3.53]. In this manner, information on the electronic band structure can also be obtained.

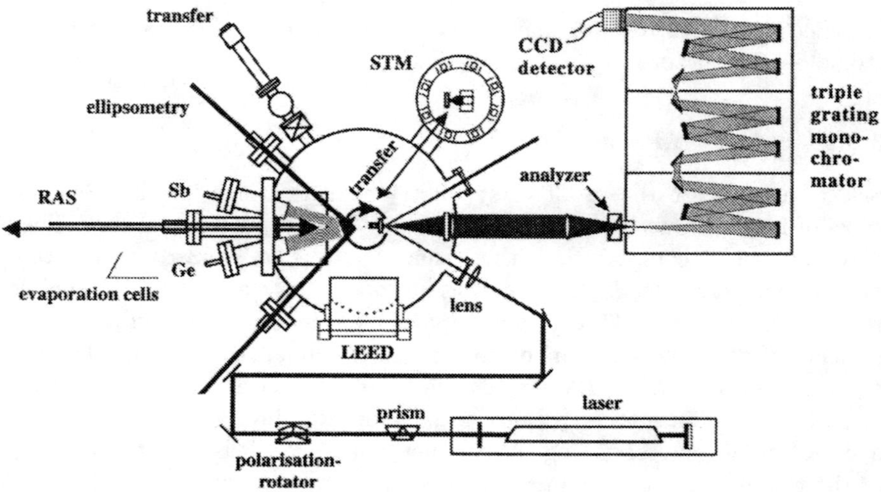

Fig. 3.2. Raman scattering setup for in situ investigations under UHV conditions. A special window arrangement optimizes the suppression of stray light while providing a largest possible solid angle of acceptance for the scattered light. Besides the Raman setup the UHV vessel is equipped with other optical windows to attach ellipsometry and Reflectance Anisotropy Spectroscopy [3.15] and with standard surface science techniques such as LEED and STM. Furthermore, a transfer mechanism for quick sample exchange and preparation facilities such as evaporation sources, sample heating and cooling (not shown) are available

3.2.5 Experimental Setup for Raman Scattering

The essential parts of a Raman setup are a laser light source, a spectrometer and a sensitive photon detector. For monitoring surface vibrations the Raman experiments must be performed while the sample is held under well defined, usually UHV, conditions. Appropriate optical viewports in the UHV vessel for the incoming and scattered light are required. A possible arrangement of a Raman spectrometer and an UHV setup is shown in Fig. 3.2.

The scattered light is collected through a lens system and focused onto the entrance slit of the spectrometer. The ray trace is adjusted such that the intense, reflected laser beam falls well outside the solid angle collected by the lens system. Furthermore, the incoming, scattered and reflected laser beams pass through separate windows, a fact which prevents the collection of stray light from the windows onto the monochromator entrance slit. This arrangement ensures an effective suppression of stray light by geometrical filtering. Special attention should be paid to a design allowing the collection of a large solid angle of scattered light. This can be achieved by the use of a re-entrant window with the collection lens placed as close as possible to the sample. Inserting a polarization analyzer in front of the entrance slit allows the selective detection of scattered light with a well-defined polarization. In combination with a polarization rotator (e.g., Fresnel rhombus) inserted in the incident laser beam (assumed to be polarized) individual components of the Raman tensor can be resolved.

For the frequency analysis of the scattered light, double grating monochromators, equipped with a photomultiplier detection system, or triple grating monochromators equipped with CCD arrays are standard. Alternatively, in case of extremely low scattering intensities, a single stage monochromator, combined with a holographic prefilter to block the intense elastically scattered light, can be used. For a very efficient reduction of the laser frequency contribution to the scattered light (Rayleigh scattering), filters have become commercially available. These so-called super-notch filters have a very steep dip in their transmission spectrum with a transmission of 10^{-6} at the centre and a spectral half width of about 150 cm^{-1}. This setup, however, is only applicable in those cases where the frequency shift of the inelastically scattered light is larger than the bandwidth of the dielectric prefilter. In many cases surface modes are expected at lower eigenfrequencies. Therefore, such type of setup is not of general use for monitoring surface vibrations. The double or triple monochromator setup, in contrast, can be applied for recording Raman lines even close to the laser line. Especially, the combination of a triple monochromator with a CCD array, which enables the parallel detection of a spectral interval with high sensitivity and low background noise [3.56, 3.57, 3.58], represents an attractive setup for investigating low intensity Raman signals.

In most cases, the frequency resolution of a Raman spectrometer is determined rather by the spectral slit width than by the diffraction limit of the gratings. Typically, a double monochromator of a focal length $f = 80\,cm$, equipped with 1800 lines/mm gratings, gives for wavelengths near 500 nm a spectral halfwidth of 1.3 cm^{-1} (0.16 meV) when the geometrical slit width is set to 100 μm. The advantage of the double monochromator/PMT setup is the possible continuous variation of spectral resolution by changing the slit width.

In the case of a monochromator equipped with a CCD detector, the width of the detector elements (pixels) limits the ultimate spectral resolution. In the case of a triple monochromator with the first two monochromators operating in the subtractive mode, the spectral width of the detected interval is determined by the focal length and the grating period of the third monochromator. For example, a monochromator with 60 cm focal length and 1800 lines/mm grating, equipped with a 1000 element detector of 2.4 cm total width, would cover, in the spectral region around 500 nm, an interval of 650 cm^{-1}, yielding a resolution of about 0.65 cm^{-1} per pixel. For high-resolution purposes some Raman triple spectrographs offer the option of switching into a triple additive mode.

Ar^+- or Kr^+- ion lasers are the most common light sources for Raman spectroscopy. They generate a number of intense, monochromatic lines from the near IR over the visible range to the UV. To optimize resonance conditions, dye lasers are often used. These lasers operate through optical pumping of a laser medium (dye) (using Ar^+ or Kr^+ lasers) which gives stimulated spectral emission in a broad interval. Apart from lasers based on liquid dyes, more recently Ti-sapphire solid-state lasers have been developed, covering the spectral range between 600 nm and 1 µm. Besides the simpler operation, an essential advantage of the Ti-sapphire laser is the long time stability over a wide power range.

3.3 Antimony Monolayers on III–V(110)

Sb on (110) surfaces of III–V semiconductors represents a prototype semiconductor interface which has been widely investigated by means of surface sensitive experimental methods and also theoretically (Table 3.1). Usually the interface between two materials possesses a certain amount of more or less disordered, difficult to characterize interfacial layer which results from chemical reaction and interdiffusion at the interface. Especially in the case of compound semiconductors, such as III–V compounds, chemical interface reactions are a very common aspect in heteroepitaxy and in metal/semiconductor interface formation [3.60, 3.61]. In the case of Sb on III–V(110) such complications are completely absent: the adsorption of Sb leads to the formation of an ideally abrupt interface on an atomic scale. Besides small modifications in atomic positions of the first few atomic layers of the III–V semiconductor (110) surfaces, the substrate remains unaffected by Sb termination [3.59]. A second important aspect concerns the relatively simple, well ordered structure of Sb monolayers on III–V(110). Since in early LEED and XPS investigations [3.62, 3.63, 3.64, 3.65, 3.66, 3.67] it was discovered that Sb monolayers terminate III–V(110) surfaces in the $p(1 \times 1)$ structure, these model systems have attracted much interest.

Along with the experimental work, the Sb-monolayer-terminated III–V(110) surfaces have attracted considerable theoretical interest, where the atomic structure and the surface electronic band structure have been calculated via various approaches. Hence, the Sb monolayer terminated III–V(110) surfaces represent most likely the best understood semiconductor interfaces at present. Table 3.1 gives an overview concerning the related experimental and theoretical work.

Because of their prototype character, the Sb monolayer terminated III–V(110) surfaces were more recently investigated by surface optical characterization techniques, i.e., Raman scattering, reflectance anisotropy spectroscopy and spectroscopic ellipsometry [3.60, 3.102, 3.103, 3.105, 3.108, 3.136, 3.137, 3.138, 3.139, 3.141, 3.142]. In surface science these optical spectroscopies represent novel techniques whose potential can be explored by applying them to well-understood systems. In the process of these studies the first evidence of Raman lines originating from surface vibrational modes was ob-

Table 3.1. Experimental and theoretical investigations reporting structural and electronic properties of Sb monolayers on III–V(110) surfaces [3.59]

Method	Property	System	References
LEED	geometry	Sb/GaAs	[3.63, 3.64, 3.68]
		Sb/InP	[3.69, 3.70]
XSW	geometry	Sb/GaP	[3.71]
		Sb/GaAs	[3.72, 3.73]
		Sb/InP	[3.73]
GIXD	geometry	Sb/GaAs	[3.74]
PED	geometry	Sb/GaP	[3.75, 3.76]
		Sb/GaAs	[3.77]
		Sb/InP	[3.75, 3.76]
		Sb/InAs	[3.78]
STM	geometry	Sb/GaAs	[3.79, 3.80, 3.81, 3.82, 3.83, 3.84]
XPS/	chemisorption	Sb/GaAs	[3.63, 3.72, 3.85, 3.86, 3.87, 3.88, 3.89, 3.90]
SXPS		Sb/InP	[3.89, 3.91]
		Sb/GaP	[3.71, 3.90]
		Sb/InAs	[3.78, 3.92]
ARPES/ARUPS	occupied	Sb/GaP	[3.90, 3.93]
	states	Sb/GaAs	[3.93, 3.94, 3.95]
		Sb/InP	[3.91, 3.96, 3.97]
		Sb/InAs	[3.92]
ARIPS	empty	Sb/GaAs	[3.98, 3.99]
	states	Sb/InP	[3.98, 3.99]
HREELS, EELS,	interband	Sb/GaP	[3.102, 3.103]
RAS, RRS,	transitions	Sb/GaAs	[3.83, 3.100, 3.101, 3.109, 3.102, 3.103, 3.104, 3.105], [3.106, 3.107, 3.141, 3.136]
ellipsometry		Sb/InP	[3.102, 3.103, 3.105, 3.110, 3.111, 3.141, 3.136]
		Sb/InAs	[3.102, 3.103]
total energy	atomic	Sb/GaP	[3.59, 3.112, 3.113, 3.115, 3.120]
minimization	geometry	Sb/GaAs	[3.59, 3.112, 3.113, 3.114, 3.115, 3.116, 3.117, 3.118] [3.119]
		Sb/InP	[3.59, 3.112, 3.113, 3.115, 3.116, 3.117, 3.118]
		Sb/InAs	[3.59, 3.112, 3.113, 3.115]
band structure	occupied	Sb/GaP	[3.59, 3.120, 3.123, 3.126, 3.121]
calculation	and	Sb/GaAs	[3.59, 3.117, 3.119, 3.122, 3.125, 3.121]
	empty	Sb/InP	[3.59, 3.117, 3.124, 3.121]
	states	Sb/InAs	[3.59, 3.121]

tained [3.137, 3.142]. This discovery has stimulated systematic research of monolayer terminated III–V(110) surfaces aimed at understanding the origin and mechanisms of the surface related Raman signals on a system where thorough theoretical modelling is available [3.59, 3.141]. Consequently, at present most of the published Raman spectroscopy data of microscopic surface vibrational modes are related to it and have led to a detailed understanding of the related Raman signals, showing that this effect is a general surface feature rather than reflecting only few especially favourable cases. In the following sections we will show that microscopic surface modes (modes confined to the immediate neighborhood of the surface) may indeed give rise to reasonable Raman signals. The pertinent scattering mechanism can be described in the deformation potential picture taking into account that surface electronic properties, distinct from those of the bulk, play the key role for inelastic light scattering from surface modes. We demonstrate that eigenfrequencies and selection rules of the surface modes confirm the results of theoretical approaches and that the large Raman cross section is a result of resonances involving electronic properties of the surface.

3.3.1 Preparation of Ordered Sb Monolayers

A prerequisite to the generation of well ordered Sb monolayers is the reproducible preparation of high-quality substrates. This requirement is guaranteed for III–V(110) surfaces by the cleaving procedure which is the most reliable and easy to handle method for the preparation of atomically clean and ordered surfaces under UHV conditions [3.61,3.62]. The preparation of Sb monolayers on III–V(110) surfaces benefits from the fact that Sb forms strong chemical bonds with the uppermost atomic layer of the substrate, whereas further Sb is weakly adsorbed on top of the first monolayer [3.59]. Therefore, during deposition of Sb on III–V(110) surfaces for submonolayer coverages, small regions covered with an Sb monolayer initially build up while some regions remain bare. The monolayer patches grow two-dimensionally with coverage until one monolayer is completed. Above monolayer coverage Sb islands occur (Stranski-Krastanov growth mode) [3.142]. By Scanning Tunneling Microscopy (STM), however, it was shown that the growth of the two-dimensional monolayer is not ideal when Sb is deposited onto III–V-substrates at room temperature. Instead, before completion of the monolayer already second layer nucleation is observed to some extent [3.79, 3.80]. The non-ideality of the as-deposited monolayer was shown to influence the electronic surface properties, giving rise to additional, localized surface states associated with atomic defects in the monolayer (missing atoms, second layer atoms) [3.79,3.80,3.85,3.86,3.89,3.142]. Fortunately, the strong bonding of the first layer to the substrate allows one to improve the monolayer homogeneity substantially: As shown by STM and Soft-X-ray Photoemission Spectroscopy (SXPS), annealing of the as-deposited Sb-layers induces ordering to an ideal monolayer structure [3.75,3.77,3.78,3.85,3.86,3.89,3.142]. The most versatile preparation procedure depends on annealing an Sb deposit of a few layers thick (in the temperature range of 200–400°C), which gives rise to the formation of a well defined, ordered monolayer through desorption of the excess Sb. This preparation procedure leads to a reproducible monolayer termination, independent of experimental limitations such as the non-ideal control of deposition thickness.

3.3.2 Structure and Electronic Properties

As stated above, the atomic structure of the Sb monolayers on III–V(110) surfaces is very well known from many experimental and theoretical investigations. Figure 3.3 shows the adsorption geometry. The structure of such Sb-monolayers is called Epitaxial Continued Layer Structure (ECLS) because the adsorption sites of the Sb adatoms correspond approximately to the positions of the next (hypothetical) III–V layer, i.e., the epitaxial Sb monolayer

a)

b)

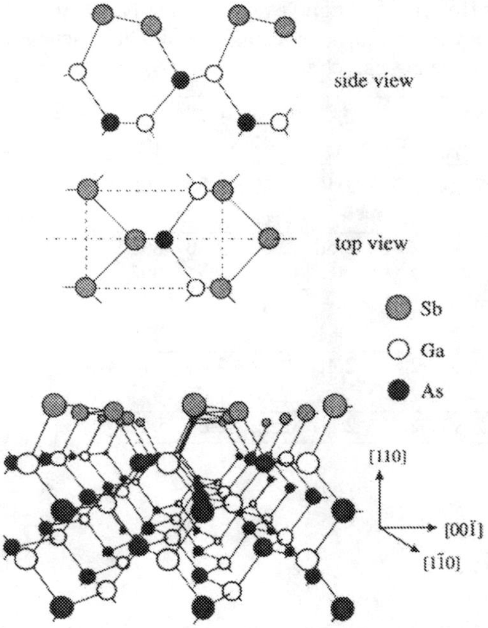

Fig. 3.3. ECLS structure of Sb monolayers. This type of structure has been found on many III–V(110) substrates, apart from Sb-containing III–Vs (GaSb, InSb), where seemingly no ordered Sb monolayer structure exists. (**a**) Side- and top-view of a surface unit mesh. The dashed-dotted line indicates the mirror plane, the only symmetry element in the C_s point group. (**b**) Perspective view on a larger scale

is a continuation of the bulk structure [3.59, 3.112, 3.113, 3.121]. The bonding configuration, distinct from that of bulk Sb, is very closely related to the bonding in III–Vs.

The atomic structure of the Sb-monolayer-terminated surface is quite similar to that of an ideal III–V(110) surface, i.e., no relaxation of the surface atoms occurs. On clean III–V(110) surfaces the ideal geometry (just cutting along a bulk plane without restructuring) is energetically unfavourable since each group-III and group-V surface atom would have one half-filled dangling bond orbital. This induces the so-called surface relaxation which corresponds to an outward motion of the group-V surface atom until a trigonal p^3 bonding configuration is reached and an inward motion of the group-III surface atom until a planar sp^2 bonding configuration is achieved. The surface relaxation is accompanied by a charge transfer from the electropositive group-III to the electronegative group-V atom. For Sb-terminated surfaces, on the other hand, the chemical and electronic inequivalence of the two surface atoms per (1×1) unit cell is lifted. Both Sb atoms have a doubly occupied non-bonding orbital already in the ideal surface geometry. Therefore, only a small relaxation, i.e., a shift outwards from the bulk atomic positions, is found for the monolayer

Table 3.2. Structural parameters (in Å) of Sb monolayers on III–V(110) substrates. For comparison, the sum of covalent radii $\sum r_{\mathrm{cov}}$ is also given [3.59]; α denotes the angle between the Sb–Sb-bonds in the zig-zag chain

		Sb–Sb	$\sum r_{\mathrm{cov}}$	Sb–V	$\sum r_{\mathrm{cov}}$	Sb–III	$\sum r_{\mathrm{cov}}$	α
Sb/GaP	ab initio [3.149]	2.78	2.80	2.55	2.46	2.59	2.66	86°
	SEXAFS [3.150]	2.88		2.60		2.79		
Sb/GaAs	ab initio [3.149]	2.81	2.80	2.66	2.60	2.59	2.66	89°
	LEED [3.68]	2.77		2.66		2.64		90°
	PED [3.77]	2.70		2.70		2.57		96°
	GIXD [3.74]	2.80						91°
Sb/InP	ab initio [3.149]	2.82	2.80	2.55	2.46	2.72	2.84	91°
	LEED [3.70]	2.82		2.52		2.80		95°
Sb/InAs	ab initio [3.149]	2.84	2.80	2.66	2.60	2.72	2.84	94°

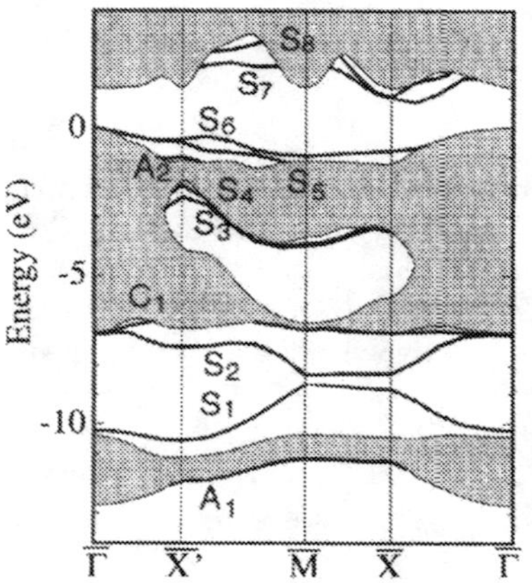

Fig. 3.4. Surface band structure for Sb-monolayer-terminated GaAs(110) superimposed on the projected bulk band structure (dotted regions) of the GaAs substrate; C_i and A_i denote states localized at cations and anions of the substrate, S_i denotes chemisorption states or states localized at the overlayer. The energy scale is referenced to the valence band maximum (0 eV) of GaAs at $\bar{\Gamma}$ [3.59]

terminated surface. The detailed atomic structure of the Sb-monolayer is somewhat dependent on the choice of III–V(110) substrate. Table 3.2 gives the Sb–Sb bond lengths in the zig-zag chain, the bond lengths between the Sb atoms and first substrate layer atoms, and the bond angle in the zig-zag chain derived by DFT-LDA total energy minimization [3.59]. The calculated bond lengths are quite close to the estimate based on the sum of the *covalent radii*, but show small, systematic variations depending on the substrate. This bond length variation is correlated with the lattice constant and the ionicity of the substrates. Thus, strain or charge transfer may affect the quantitative details of the ECLS geometry [3.59].

The electronic properties of the Sb-monolayer terminated surfaces have been studied in numerous band structure calculations (Table 3.1). We summarize the most important results because of their great importance in understanding of Raman scattering by surface phonons (Sect. 3.3.5). Due to the similar adsorption geometry and chemical bonding the main features of the surface electronic band structure are rather common for the different III–V(110) substrates. Figure 3.4 shows, as an example, the result of LDA-DFT calculations for Sb on GaAs(110). Very similar surface bands have been derived in earlier tight binding calculations [3.121]. Besides the shaded regions corresponding to the projected bulk band structure, the thick solid lines represent surface electronic bands, i.e., states which do not overlap with bulk electronic states. The energy scale is referenced to the top of the bulk valence band, i.e., electronic states below 0 eV are occupied, those above are empty. The notation of the surface bands is adapted from earlier pseudopotential calculations [3.122]. An analysis of the orbital character of the surface states was performed on the basis of tight binding calculations [3.121]. The orbital character of these states derived on the basis of the early tight binding calculations [3.121] agrees with the results of recent LDA-DFT investigations [3.59]. Anion (group-V atom), cation (group-III atom) and Sb related surface states are distinguished by the notation A_n, C_n and S_n, respectively. Most relevant to the optical studies are the states S_5, S_6 (filled) and S_7, S_8 (empty) which contribute to optical excitations in the visible/near UV range. According to these studies, the S_5, S_6 (filled) and S_7, S_8 (empty) states are bonding and antibonding states formed between Sb chain atoms and underlying substrate atoms. The pair S_5 and S_7 belongs to the Sb–group-V-atom bonding, the pair S_6 and S_8 to the Sb–group-III-atom bonding [3.59,3.121]. These orbitals have p_z character with contributions of substrate atom orbitals and Sb atoms, forming a strong σ bonding [3.59,3.121]. Figure 3.5 shows the charge density distribution of the surface states at special point of the SBZ derived from the DFT-LDA calculation. The plot clearly reveals the confinement of the electronic states to the first few atomic layers. As earlier suggested on the basis of the tight binding calculations [3.121], the charge density plot confirms the different orbital characters of the surface states due to contributions of either anion or cation substrate atoms.

These states are of possible importance for the interaction with photons, treated in the following sections, since the energy separation of occupied A_2, S_5, S_6 to unoccupied S_7, S_8 states falls into the relevant energy regime.

3.3.3 Surface Phonons

The surface vibrational properties of Sb monolayer terminated III–V(110) surfaces have been probed by Raman spectroscopy for the cases of GaAs, InP, GaP, InAs and InSb substrates [3.103, 3.105, 3.108, 3.136, 3.137, 3.138, 3.139, 3.141, 3.142]. In all cases except that of the InSb substrate, vibrational modes related to the Sb monolayer were detected. This result has been corroborated in other studies revealing that in case of InSb substrates the ordered (ECLS type) monolayer, whose vibrational properties give rise to the so-called *monolayer modes* in the Raman spectra, is not formed [3.59]. The

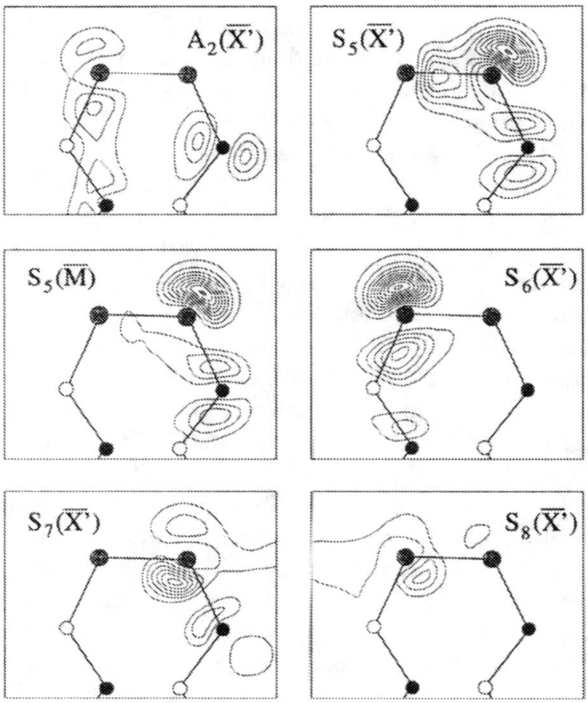

Fig. 3.5. Contour plots of charge densities associated with the bound surface states for Sb-monolayer-terminated GaAs(110). The A_2, S_5, S_6, S_7, and S_8 states (relevant for optical transitions) are shown at the $\bar{X}'$ and $\bar{M}$ points of the SBZ [3.59]

most complete experimental study of the monolayer modes has been performed for the Sb/InP(110) system, where monolayer Raman signals were detected for the first time [3.137, 3.142]. Figure 3.6 shows a sequence of Raman spectra recorded from InP(110) surfaces under UHV conditions after various deposition steps of Sb (nominal total coverage of 0 to 1.0 ML).

On the clean surface, Raman lines due to the transverse (TO) and longitudinal (LO) optical phonon modes of the bulk InP show up. Moreover, one to two orders of magnitude weaker features appear in the frequency range 70–200 cm^{-1} and are related to second order phonon scattering [3.127, 3.128, 3.129] (see Sect. 3.3.4). With increasing coverage three intense Raman lines, in the frequency range of 70–200 cm^{-1}, and three smaller ones in the bulk phonon range marked by dash-dotted lines develop. They are associated with the surface vibrational modes of the Sb monolayer. The experimental linewidth (full width at half maximum) of these modes is 3 cm^{-1} (0.4 meV), limited by the monochromator resolution in these experiments. The assignment of these lines to surface modes is further supported by the increase in scattering intensities with increasing Sb coverage up to 1ML (Fig. 3.7).

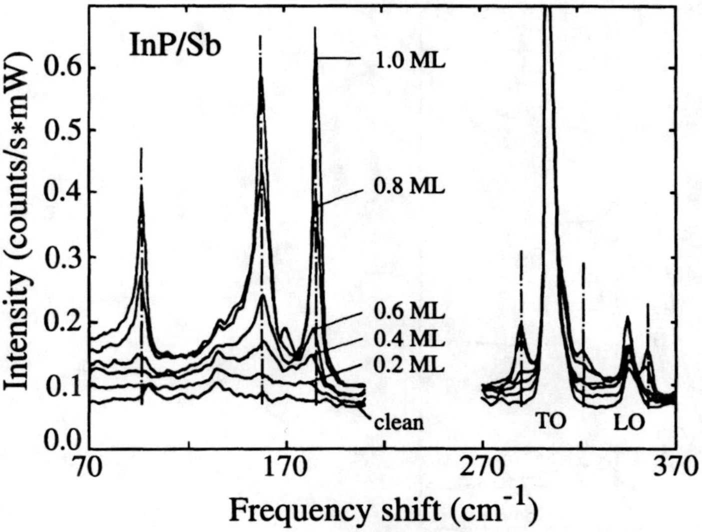

Fig. 3.6. Raman spectra of Sb/InP(110) for coverage of 0–1.0 ML. The Raman experiments were performed using the 514.5nm (2.41eV) line of an Ar-ion laser. The scattering geometry (in the PORTO notation, see Sect. 3.2.4) was $[1\bar{1}0]([1\bar{1}0];[1\bar{1}0])[110]$ (surface normal along [110]) [3.137]. Both spectral intervals are shown on the same intensity scale

The development of the peak intensities with coverage shows that the respective modes are correlated with the degree of completion of the monolayer. This result is expected since according to the finding of STM investigations and LDA-DFT based total energy calculations [3.59,3.79,3.80,3.119] for sub-monolayer coverage patches of regular ECLS type covered regions form which grow two-dimensionally up to the full ML coverage. The persisting monolayer peaks for coverage exceeding 1ML demonstrate that the Sb monolayer, due to its strong bonding to the substrate, remains essentially intact under the layer deposited on top [3.137, 3.140]. In fact STM experiments performed on Sb multilayers deposited on III–V(110) substrates have shown that the amorphous excess Sb is only weakly bonded to the monolayer-terminated substrate [3.143, 3.144].

The initial observation of surface phonon modes of the Sb-monolayer terminated InP(110) surface stimulated a more thorough analysis of selection rules and resonance of the Raman cross section. This work is summarized in the next two sections.

3.3.4 Symmetry Considerations and Selection Rules

The monolayer modes shown in Fig. 3.6 represent only part of the surface vibrational eigenmodes observed in the Raman spectra. This is due to the Raman selection rules: by choosing the polarization of incident and scattered

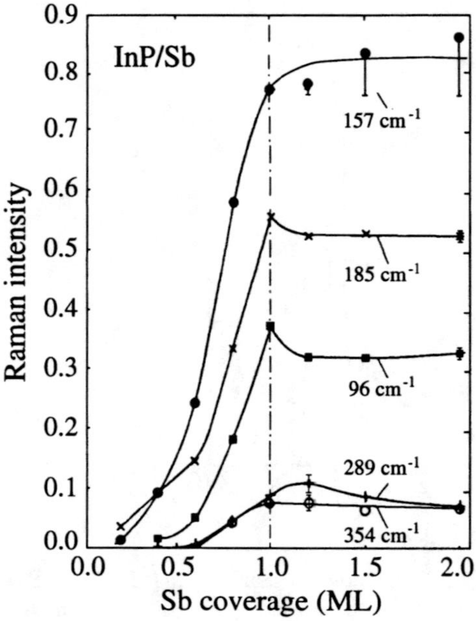

Fig. 3.7. Intensity of the monolayer vibrational modes normalized to the TO phonon scattering intensity as a function of Sb coverage. After completion of the monolayer, the Raman lines remain visible with nearly constant intensity. The weak increase in intensity of the 157 cm^{-1} mode after completion of 1ML is due to an overlap with an Sb bulk phonon mode [3.137]

light along the $[1\bar{1}0]$-direction of the surface, only modes having a nonzero diagonal component in the respective Raman tensor are sampled. The complete analysis for different combinations of incident and scattered light polarizations is shown in Fig. 3.8 for Sb monolayer covered InP(110) (a) and GaAs(110) (b). In contrast to Fig. 3.6 the monolayer structures were prepared by deposition of 4 ML Sb and subsequent annealing to 550 K inducing the desorption of the excess Sb and ordering of the monolayer structure [3.142]. The top spectrum of Fig. 3.8a was measured in the same polarization configuration as those in Fig. 3.6, taken after deposition without subsequent annealing. No significant difference between the two monolayer spectra is observed, showing that the surface lattice dynamics is not noticeably influenced by a small defect density, as is the case in the monolayer structure before annealing. The surface electronic structure, in contrast, has been shown to be influenced more sensitively by the degree of order [3.89, 3.105].

Comparing InP Fig. 3.8(a) with the GaAs (b) case, the most evident difference is the approximately four times larger scattering intensity of the monolayer modes of InP. Apart from this, a strong dependence on polarization is evident. For the bulk LO and TO phonon modes the well-known group theory selection rules corresponding to the point group T_d are obeyed [3.2, 3.55, 3.60].

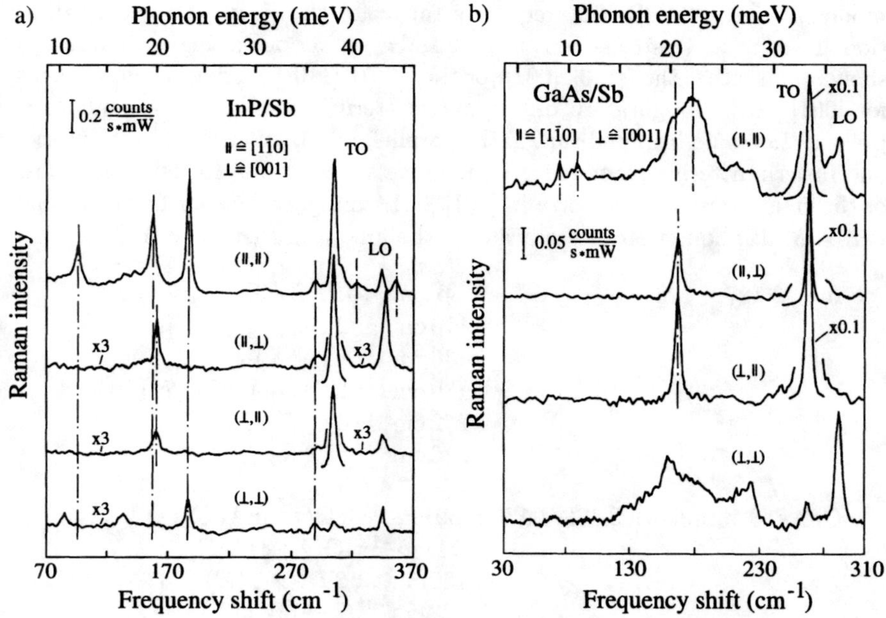

Fig. 3.8. Raman spectra of ML-terminated InP(110) (**a**) and GaAs(110) (**b**) for different polarization configurations $[\bar{1}\bar{1}0](x,y)[110]$, with x,y parallel or perpendicular to the Sb chain direction along $[1\bar{1}0]$ [3.60,3.139]. Monolayer vibrational modes are marked by dashed-dotted lines. All spectra were recorded using the 2.41 eV line of an Ar laser. For GaAs the intensity scale is expanded by a factor of four with respect to InP (note the scale bars inside the figures). The energy scale is given in meV and cm^{-1}, $1\,\mathrm{meV} = 8.065\,\mathrm{cm}^{-1}$

Raman scattering involving the Deformation Potential (DP) mechanism is observed from the TO phonon for polarizations of incident and scattered light $e_{\mathrm{i}}, e_{\mathrm{s}}$ along ($[1\bar{1}0], [1\bar{1}0]$); ($[1\bar{1}0], [001]$); and ($[001], [1\bar{1}0]$). In the ($[001], [001]$) configuration, TO phonon scattering is symmetry forbidden. Deformation potential scattering from the LO mode is symmetry forbidden in all polarization configurations, but shows up through the Fröhlich (F) scattering mechanism for parallel polarization configurations, i.e., ($[1\bar{1}0], [1\bar{1}0]$) and ($[001], [001]$). Moreover, a small DP contribution of the scattering by LO phonons arises due to the non-ideal backscattering geometry. The deformation potential and F scattering from bulk zincblende phonons has been treated extensively in many previous articles e.g., [3.2, 3.14, 3.60].

Besides the bulk modes, the monolayer vibrational modes also show a pronounced dependence on the choice of polarization directions. The selection rules reflect the fact that the two-dimensional monolayer structure, Fig. 3.3, is highly anisotropic: Sb atoms are adsorbed in zig-zag chains extending along the $[1\bar{1}0]$- direction, not linked in the $[001]$-direction. For Sb on InP(110) three strong modes are observed in the lower frequency range and three weak modes in the bulk LO–TO frequency range for the ($[1\bar{1}0], [1\bar{1}0]$) configuration. Two modes (a strong one at lower and a weaker one at higher frequency) show up for either the ($[1\bar{1}0], [001]$) or the ($[001], [1\bar{1}0]$) configuration, whereas for [001],[001] configuration only weak scattering from monolayer modes is present. It should be noted that in the parallel polarization configurations second order scattering from acoustic phonons arises; it is responsible for most of the structures of the spectra in [001],[001] configurations for both InP and GaAs. Multiphonon processes occur at the following frequency shifts:

InP [3.127, 3.128]:	86 cm^{-1}: LO(L)–TO(L)
	110 cm^{-1}: 2TA(L)
	136 cm^{-1}: 2TA(X)
	189 cm^{-1}: 2TA(W), 2TA(K)
	252 cm^{-1}: TO(X)–TA(X)
	276 cm^{-1}: LO(X)–TA(X)
GaAs [3.130, 3.131, 3.132, 3.133]:	52 cm^{-1}: TO(L)–LA(L)
	71 cm^{-1}: IO(Σ)–IIA(Σ)
	160 cm^{-1}: 2TA(X),(Σ)
	223 cm^{-1}: 2TA(W),(Σ)

For Sb on GaAs(111), in contrast to the InP case, the Raman signals corresponding to monolayer vibrational modes in parallel configuration are less pronounced. The scattering intensities have the same order of magnitude as the acoustic multiphonon features. By comparison with the respective spectra of the clean GaAs surface four weak structures and a broad feature possibly containing two additional lines are identified in the ($[1\bar{1}0]; [1\bar{1}0]$) configuration, one strong feature is clearly present for crossed polarizations, and no significant contribution of the monolayer modes shows up in the [001],[001]-configuration.

The eigenfrequencies of the monolayer vibrational modes are summarized in Table 3.3. Besides InP and GaAs, Raman results obtained from Sb monolayer terminated GaP(110) and InAs(110) are also included. Phonon modes observed in parallel configuration are labelled A', in crossed polarization configuration A''. This notation reflects the group theory assignments discussed below.

If the surface induced Raman signals are due to symmetry-allowed first-order Raman scattering from surface optical modes, they must obey well-defined selection rules dependent on the symmetry of the structure (point group) (Sect. 3.2.4) [3.2, 3.14]. This question will be addressed in the following. The number of observed surface phonon modes amounts to 6 (4) A'- and 2 (1) A''-modes for the InP (GaAs) substrates, respectively. In order to describe 8 optical phonon modes, in the case of a three-dimensional infinite translation lattice, a four-atom base is required. Then 12 modes are obtained, three of which correspond to acoustic and 9 to optical branches.

The Sb monolayer system has a two-dimensional translation lattice. Considering only the outermost atomic layer of Sb, the lattice has a two-atom base (Fig. 3.3). However, since translation invariance along the surface normal is broken, the atoms of the underlying substrate layers all belong to the atomic base. The number of atoms in the base thus depends on the number of layers taken into account: the more layers considered, the more phonon branches arise, branches that may represent surface- as well as bulk-like modes. To describe the experimentally observed surface modes at least a four-atom base corresponding to two layers (Sb monolayer and the upper substrate layer) has to be taken into account.

The only symmetry operation which applies to the Sb monolayer, besides the translations of the two-dimensional lattice, is a mirror plane, as indicated in Fig. 3.3. The monolayer structure thus belongs to the point group C_s. According to the number of irreducible representations given in [3.2, 3.55], 6 non-degenerate optical branches of A' and three non-degenerate of A'' symmetry should appear in the phonon dispersion relations. Taking into account that the polarization directions of the incident and scattered light are oriented within the two-dimensional surface plane, the Raman tensors for the A'- and A''–modes (in coordinate system $x' = [001], y' = [1\bar{1}0], z' = [\bar{1}\bar{1}0]$ are [3.2]:

$$\tilde{R}(A') = \begin{pmatrix} a & 0 \\ 0 & b \end{pmatrix} , \ \tilde{R}(A'') = \begin{pmatrix} 0 & c \\ c & 0 \end{pmatrix} . \tag{3.11}$$

Consequently, first-order Raman scattering by the $6A'$–modes should take place only for parallel polarizations and by the $3A''$–modes in crossed polarizations (we assume that the polarizations are along the x'- or y'-axes). This group theoretical prediction agrees very well with the experimental findings, thus supporting the interpretation given above based on first order deformation potential scattering. However, the low scattering efficiency for the A'–modes in ([001],[001]) polarization configuration remains unexplained. This effect reflects the magnitude of the tensor components a and b responsible for scattering in the ($[1\bar{1}0], [1\bar{1}0]$) and ([001],[001])-configurations, respectively, as we will see in the following section where a quantitative determination of the Raman efficiency is presented.

The total number and symmetry of the possible eigenmodes has been determined by group theory arguments, but the normal coordinates of the respective modes are still unknown. Fortunately, the lattice dynamics of Sb monolayers on III–V(110) was thoroughly investigated within a *frozen-phonon* approach based on a LDA-DFT total energy calculation scheme [3.59]. In this work the upper three atomic layers (Sb layer plus two substrate layers) of a 10-layer slab were allowed to move, while the other layers were fixed in the bulk geometry. More recently the full lattice dynamics of such slabs has been calculated within the Density Functional Perturbation Theory (DFPT) approach [3.134]. The eigenfrequencies derived from these calculations are also included in Table 3.3.

Apart from one exception, the $4A'$–mode of Sb/InP, the eigenfrequencies calculated with the *frozen-phonon* approach compare quite well with the experimental results. The DFPT calculations, however, reproduce the experimental results with significantly higher precision, in particular also for those modes, like the $4A'$–mode of Sb/InP(110), where the frozen-phonon calculations fail. As discussed below in more detail this finding demonstrates that the surface vibrational modes partly couple with bulk phonon excitations. The eigenvectors of the modes calculated in the *frozen-phonon* approach are shown in Fig. 3.9 for the InP substrate as an example, to give a plausible picture of the surface vibrational modes. It should be noted that the eigenvectors depend on the choice of III–V(110) substrate due to the different mass of the substrate atoms [3.59]. The atomic displacements of the A''–modes are along the $[1\bar{1}0]$-direction, while those of the A'–modes are in the $(1\bar{1}0)$-plane. Since the mirror plane corresponds to a $(1\bar{1}0)$-plane, the A'' displacements lift the mirror symmetry while it is preserved by the A' displacements.

The displacement patterns already indicate that the surface eigenmodes are not necessarily confined to a vibration of the top few atomic layers. The eigenfrequencies of the monolayer vibrations in fact are found in a spectral region where they overlap with the one-phonon density of bulk states of the respective substrate materials. Figure 3.10 shows the dispersion curves of the bulk phonon states of InP and GaAs, together with the eigenfrequencies of the Raman modes for Sb/InP(110) and Sb/GaAs(110), respectively. Due to the spectral overlap, a coupling between surface–bulk modes is possible. This coupling is only partially accounted for by the frozen-phonon calculations since in these calculations, for computational reasons, only the topmost three atomic layers are free to vibrate, whereas the underlying ones are fixed.

Table 3.3. Energies (in meV, 1 meV = 8.065 cm^{-1}) of the surface phonon modes observed by Raman spectroscopy on Sb-monolayer-terminated III–V(110) surfaces, numbered with increasing eigenfrequency. For comparison, the respective calculated eigenenergies for the ECLS structure at $\bar{\Gamma}$ are included. DFT refers to frozen phonon, DFPT to perturbation theory calculations

	GaP/Sb			InP/Sb			GaAs/Sb			InAs/Sb		
	RS [3.137]	DFT [3.149]	DFPT [3.134], [3.214]	RS [3.137]	DFT [3.149]	DFPT [3.134]	RS [3.138]	DFT [3.149]	DFPT [3.134]	RS [3.103]	DFT [3.149]	DFPT [3.134]
$1A'$		11.1	10.7	11.9	11.0	12.6	9.2	9.8	9.5		7.5	
$2A'$		13.1		19.5	17.7	19.6	11.0	11.1		11.9	10.3	11.4
$3A'$		21.9	22.7	22.9	20.9	23.0	21.0	23.2			20.2	18.6
$4A'$		26.6	25.0	35.8	24.0	35.8	22.5	24.1	22.4	21.6	22.5	21.7
$5A'$		39.7	39.0	39.8	37.5	40.8		27.2	26.6	24.5	25.1	24.5
$6A'$			44.8	43.9	47.6	44.7		34.6			31.3	29.9
$7A'$		51.3	48.5									
$1A''$		8.4			8.4	8.1		7.8			6.9	6.8
$2A''$	20.7	20.8	20.9	20.0	19.6	19.4	20.6	19.9	19.7	19.1	19.0	18.7
$3A''$			41.6	36.0	40.3	35.5		30.5	29.3			24.6
$4A''$		44.0	44.6								27.4	26.9

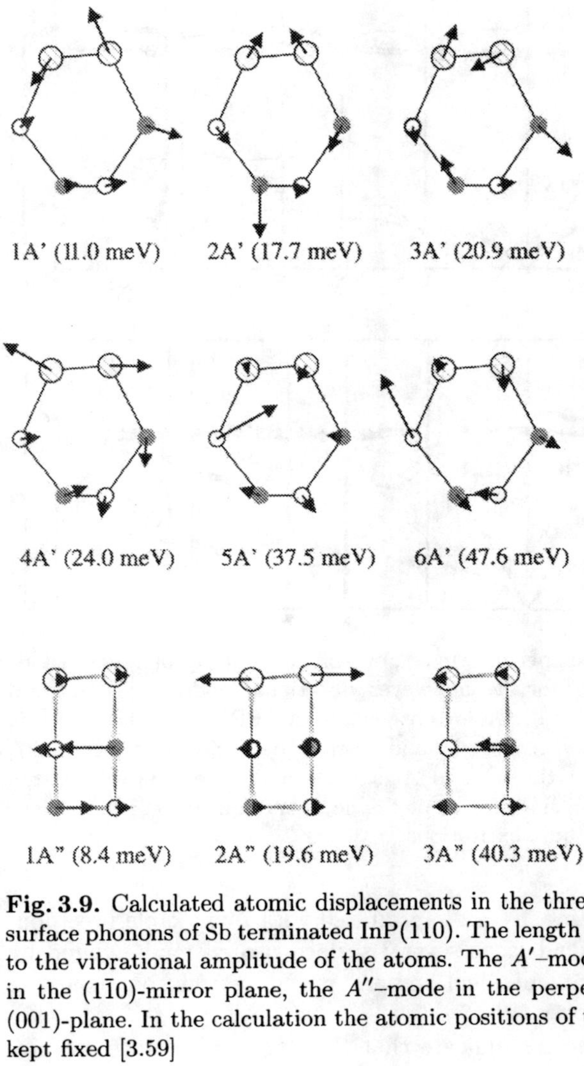

Fig. 3.9. Calculated atomic displacements in the three topmost atomic layers for surface phonons of Sb terminated InP(110). The length of the arrows is proportional to the vibrational amplitude of the atoms. The A'–modes are shown as a side-view in the $(1\bar{1}0)$-mirror plane, the A''–mode in the perpendicular direction, i.e., the (001)-plane. In the calculation the atomic positions of the fourth atomic layer were kept fixed [3.59]

Consequently, a systematic deviation of calculated and experimental eigenfrequencies should occur in the case of strong surface–bulk coupling. In particular, this accounts for the large deviation between experimental and theoretical eigenfrequency of the $4A'$–mode for Sb/InP (Table 3.3). The eigenfrequency of the $4A'$–mode is located only slightly below the dispersion branch of the bulk TO–phonon mode, whereas the calculated result yields an eigenmode at the upper border of the acoustic branch, coupled with the LA(X) bulk mode. This discrepancy is lifted in the DFPT approach where the full dynamical solution of a slab is considered. Then the mode frequencies, surface localized

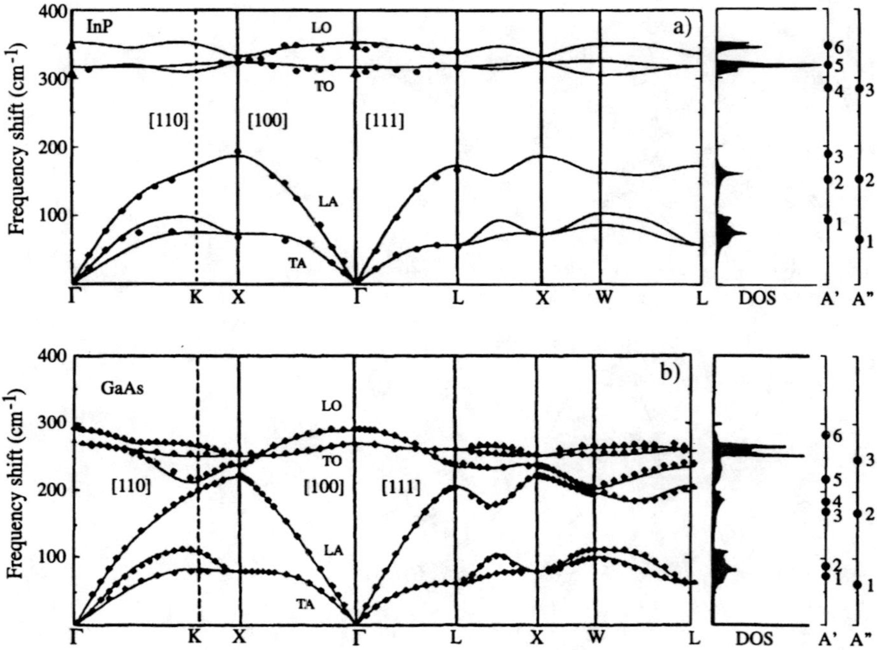

Fig. 3.10a,b. Bulk phonon dispersion curves, one-phonon density of states (DOS) and Raman shifts determined for the monolayer vibrational modes of InP (**a**) and GaAs (**b**). Dispersion curves and DOS are obtained from DFPT calculations [3.135, 3.134]. Data points correspond to neutron and Raman spectroscopy results [3.127, 3.132, 3.133]. Raman data for the A'- and A''-surface-modes are indicated on the extra scales on the right-hand side of each plot. The eigenfrequencies of the surface modes overlap with those of bulk excitations

as well as coupled to bulk modes, are reproduced with remarkable precision. However, in the DFPT calculations several surface related modes found by Raman spectroscopy can not be identified any more due to strong coupling with bulk modes.

The calculated eigenvectors indicate that the degree of surface confinement may differ for the various modes. In the case of Sb/InP(110), three other modes besides the $4A'$ fall into the frequency region of the TO and LO bulk phonon dispersion branches, the $5A'$, $6A'$ and the $3A''$–modes. The $3A''$–mode, for instance, is a localized vibration of the first substrate layer which does not involve a vibration of the atoms in the Sb monolayer. The $5A'$– and $6A'$–modes, on the other hand, do not show a significant vibrational amplitude in the topmost (Sb) layer. Therefore, in particular the A'–modes overlapping with the bulk optical phonon branches show a strong coupling of surface to bulk modes [3.30, 3.59].

The low-frequency modes, however, may couple to acoustic branches of the bulk phonon spectrum. In fact the eigenfrequencies of the $2A'$–, $3A'$–

and $2A''$–modes overlap with the LA branch, those of the $1A'$– and $1A''$–modes with the TA branch. However, in contrast to the higher-energy A'–modes coupled with optical branches, the eigenvectors of all low frequency modes involve considerable displacements of the Sb monolayer atoms (as expected since Sb is a relatively heavy atom). The $3A'$–mode coincides with the LA frequency at the X–K region of the Brillouin zone where a large density of one phonon bulk states exists, indicating possibly a more significant coupling to bulk states. Thus the calculations of the surface modes suggest that surface dynamical properties are essential for the $1A'$–, $2A'$–, and $3A'$–modes, although the latter may couple to the bulk LA(X) modes. In contrast to the cases discussed so far, the $2A''$–mode is confined to a vibration of the surface Sb atoms. The strong surface confinement of the $2A''$–mode can be understood from its atomic displacement pattern. Due to the opposite displacements of the Sb surface atoms along the $[1\bar{1}0]$-direction, the net forces acting on the underlying layers compensate each other to a large extent.

For the DFPT results, on the other hand, the agreement with the Raman frequencies is almost perfect for all modes, independent of the degree of surface confinement. Consequently, the real eigenvectors may differ significantly from the frozen-phonon results for some of the modes. The dominant surface character of the low frequency modes, however, is consistently indicated by the coincidence of eigenfrequencies derived by the different methods as well as by the resonance of the Raman cross section described in the following section.

The overview given in Table 3.3 reveals that in most cases (besides the favorite Sb/InP(110)-system) only some of the theoretically predicted surface modes are found experimentally. This is due to the fact that the Raman scattering intensity of the surface modes depends on electronic surface properties and on the normal vectors of the individual modes. Weak modes are hard to identify because their Raman intensity may be either below the instrumental detection limit or hidden by superposition on a much stronger background from the second-order bulk phonon spectrum.

In Sect. 3.1 it was stated that the microscopic optical surface phonon modes should depend rather critically on the atomic structure of the surface. Therefore, the comparison of experimentally determined mode frequencies and corresponding calculations for different, competing structure models should be a sensitive probe for determining the actual surface geometry. For Sb monolayers on GaAs(110), for instance, several different structural models were initially proposed [3.59, 3.62, 3.115]. Besides the finally established Epitaxial Continued Layer Structure (ECLS) until recently the so-called Epitaxial On Top Structure (EOTS) [3.59, 3.115] was found to well describe the available experimental data. Therefore, the hypothetical surface phonon modes for the EOTS have also been calculated for Sb on GaAs(110), in order to demonstrate the capability of distinguishing various possible surface geometries on the basis of the observed mode frequencies. For the EOTS the calculated eigenfrequencies are 13.1 meV ($1A'$), 15.0 meV ($2A'$), 21.6 meV ($3A'$), and 17.9 meV ($2A''$). The $3A'$–mode at 21.6 meV cannot be used to discriminate between the two cases since it falls into a region where a rather broad spectral band is observed experimentally (Fig. 3.8b). Comparing the calculated eigenfrequencies of the other modes with the experimental values, it turns out that in the case of the ECLS all mode frequencies agree within small deviations of less than 1 meV, while in the case of the EOTS discrepancies of 3 meV or larger arise. Thus, the comparison clearly favors the ECLS over the EOTS structure [3.151].

3.3.5 Raman Scattering Efficiency

At first glance, one of the most surprising results of the experimental observation of surface vibrational modes by Raman scattering is the unexpectedly large scattering intensity. For Sb/InP(110), the intensity of the low-frequency surface modes for 2.41 eV excitation energy is about 1/3 of that of the symmetry allowed bulk TO phonon mode. According to (3.6) in Sect. 3.2.2 the relative Raman intensities should be determined by the scattering length under the assumption of approximately similar Raman scattering cross sections for surface and bulk modes, respectively. The scattering length for the bulk modes is limited by the penetration depth of light, which amounts to 90 nm in InP at the 2.41 eV photon energy used for the Raman experiments [3.40]. For the surface modes the scattering length is limited by their confinement to a few atomic layers, predicting a surface signal below 1/100 of that of the bulk TO phonon. The Raman cross section, however, may vary quite strongly due to resonances with the intermediate electronic states. The high cross section observed for the surface phonons thus points to the fact that distinct electronic states participate in the Raman scattering processes involving surface and bulk phonons. In order to confirm this conjecture, and as a first step towards a microscopic understanding of the Raman scattering process from surface vibrational modes, the resonance of the Raman cross section was analyzed experimentally for Sb monolayer terminated InP, GaAs and InAs(110) [3.103, 3.105, 3.138, 3.139, 3.141]. The most complete characterization was performed for the intense vibrational peaks of an Sb monolayer on InP(110). Figure 3.11 shows a plot of corresponding Raman spectra recorded for several photon energies of the exciting laser beam.

For a quantitative analysis one has to normalize the measured Raman scattering intensity according to [3.2, 3.103]:

$$I_{\mathrm{s}} = A(\omega)\,[1 - R(\omega)]^2\, L\,[n_{\mathrm{BE}}(T) + 1]\,\omega^3 |\mathbf{e}_{\mathrm{s}} \tilde{R}^j \mathbf{e}_{\mathrm{i}}|^2 I_{\mathrm{i}}\ , \qquad (3.12)$$

where I_{s} expresses the scattered intensity in photon counts, I_{i} the intensity of the incident laser light, n_{BE} is the Bose-Einstein occupation probability, $\mathbf{e}_{\mathrm{i,s}}$ the polarization of incident and scattered light, $\tilde{R}^j$ the Raman tensor of a phonon mode labeled j, ω the light frequency, $A(\omega)$ the throughput

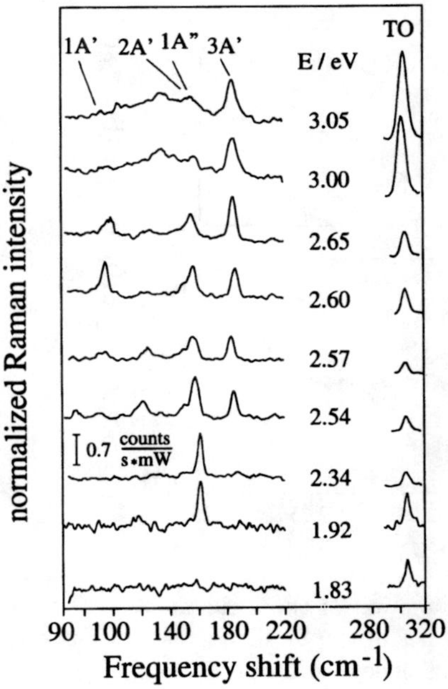

Fig. 3.11. Raman spectra of 1ML Sb on InP(110) recorded for different excitation photon energies as indicated in the figure. The incident light polarization was along $[1\bar{1}0]$; the scattered light is unpolarized. Beside the monolayer vibrational modes, the InP TO-phonon peak ($304\,\mathrm{cm}^{-1}$) also occurs [3.139]. The scattering intensity in the TO range has been scaled by a factor of 0.1

of the apparatus, L the depth characteristic for the scattering volume and $R(\omega)$ the reflectivity of the sample. Equation (3.12) follows from that given by *Cardona* [3.2] if I_s is expressed in photon count rate. In the case of Raman scattering from bulk excitations the characteristic length L is usually determined by light absorption. In the case of confined surface phonons one would have to consider the number of atomic layers involved in the mechanical vibration. Since L is difficult to pinpoint for surface phonons we use the product $L \cdot |e_s \tilde{R}^j e_i|^2$ as the surface scattering efficiency. The Raman spectra shown in Fig. 3.11 have been corrected for the effect of the photon energy dependent light throughput of the detection system (monochromator and light detector), determined separately by means of a calibrated tungsten ribbon lamp, and also for the reflectivity of the sample (optical properties are known from spectroscopic ellipsometry data [3.15, 3.16, 3.40]). Since the TO-phonon peak of the InP substrate was also recorded the correction of the Raman intensities was cross-checked against previously published data, showing an excellent agreement of the absolute TO-phonon Raman efficiency [3.51]. From the plot of corrected spectra (Fig. 3.11) it is immediately evident that the resonance behaviour of the monolayer vibrational modes is distinct from that of the bulk TO phonon, thus substantiating the initial suggestion that other electronic states are involved in these scattering processes. Furthermore, the spectra demonstrate that the resonance of the monolayer vibrational modes is different for the various eigenmodes observed. A more detailed quantitative picture of the Raman scattering efficiency of the monolayer vibrational modes as a function of incident photon energy is shown in Fig. 3.12.

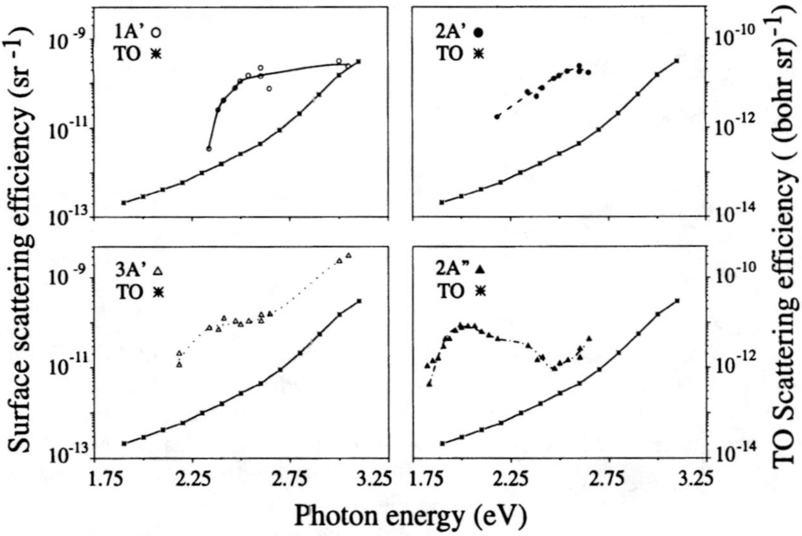

Fig. 3.12. Raman scattering efficiency as a function of the incident laser photon energy for the different surface vibrational modes of the Sb monolayer (in units of sr^{-1}). For comparison, the InP TO phonon efficiency [in units of $(\text{bohr} \cdot \text{sr})^{-1}$] is also shown [3.60]

The photon energies at which the monolayer resonances occur are clearly different from those of the bulk TO-phonon. The $2A''$-surface-mode shows a pronounced resonance around 2 eV photon energy, the $1A'$– and $2A'$–modes at around 2.5–2.7 eV, and the $3A'$–mode at 2.5 eV and also around 3 eV. The bulk TO-phonon mode shows its well known resonance profile of increasing scattering efficiency when approaching the E_1 (3.14 eV) and $E_1 + \Delta_1$ gap (3.24 eV) of InP [3.51, 3.145]. The distinct resonance behaviour of the surface phonons has been interpreted in terms of resonance with surface electronic states related to the monolayer terminated surface [3.138, 3.139]. This explanation is reasonable since electronic surface transitions attributed to empty and filled states of the monolayer-terminated surfaces have been found by other techniques around 2.1 eV and 2.65 eV [3.59, 3.102, 3.108, 3.136, 3.141]. Moreover, the band structure calculations discussed in Sect. 3.3.2 (see for instance Fig. 3.4) reveal surface bands related to the Sb–substrate bonds appearing just below the valence band maximum (filled) and above the conduction band minimum (empty). These states would explain the experimentally determined surface transitions at 2 eV and 2.5–2.7 eV. Combining the information about the spectrum of surface states and the Raman resonance profile of the monolayer modes we conclude that Raman scattering from monolayer modes is related to surface electronic states. They are specific to the Sb-monolayer-terminated surface, i.e., originate from the first few atomic layers where the bonding differs from that of the III–V bulk. Taking into account both the selection rules discussed in Sect. 3.3.4 and the resonance profile, the Raman scattering from monolayer vibrational modes is related to deformation potential scattering of surface modes modulating surface electronic states (bonds). Moreover, the fact that the resonance profile of the surface modes is clearly distinct from that of the bulk modes, i.e., not dominated by the bulk electronic band structure, shows that at least the four modes $1A'$, $2A'$, $3A'$, and $2A''$ discussed here must have a predominant surface vibrational character, in spite of some possible coupling to bulk modes. The $3A'$–mode probably exhibits the strongest coupling to bulk phonons indicated by the increasing Raman efficiency when approaching the E_1-gap of the InP substrate. These conclusions support the discussion in Sect. 3.3.4 based on the frozen-phonon approach where the individual modes are shown to possess different atomic displacements.

In the deformation potential picture the Raman scattering cross section should be related to the dielectric properties of the material by (3.8) and (3.9). As demonstrated for resonance Raman scattering from bulk semiconductors, the resonance profile can be approximated using the dielectric function of the material according to (3.10) [3.2,3.14,3.52]. In a similar way, it should be possible to test the suggested relationship for the Raman scattering from surface vibrational modes to surface electronic properties. For this purpose the contribution of the monolayer terminated surface to the dielectric response must be determined.

A well-established experimental probe of the dielectric properties of solids is Spectroscopic Ellipsometry (SE) [3.15, 3.16, 3.40]. Due to the large penetration depth of light, the dielectric function determined in an ellipsometry experiment represents an average of a bulk contribution and a surface contribution, referred to as the pseudo-dielectric function. However, despite the fact that the surface contribution is usually small compared to that of the bulk, it has been demonstrated that it can be separated with high precision [3.100, 3.101, 3.136, 3.146, 3.148]. In order to determine surface dielectric properties, as represented by the so-called *surface excess function* approximation [3.146], two ellipsometry experiments on a sample for different surface conditions, i.e., clean and adsorbate covered, have to be performed. The surface excess function derived from the difference of two pseudo-dielectric functions is a measure of the surface contribution and depends on surface conditions, unlike the constant bulk contribution. Alternatively, surface dielectric properties may be analyzed by the Reflection Anisotropy Spectroscopy (RAS) method, also known as Reflection Difference Spectroscopy (RDS). In this technique, the difference in normal incidence reflectance for two perpendicular polarization directions is measured. For cubic materials such as the III–V semiconductors the bulk related optical anisotropy in the opaque frequency regime is very weak. Thus, this technique represents a surface sensitive optical probe [3.15, 3.16, 3.40, 3.108].

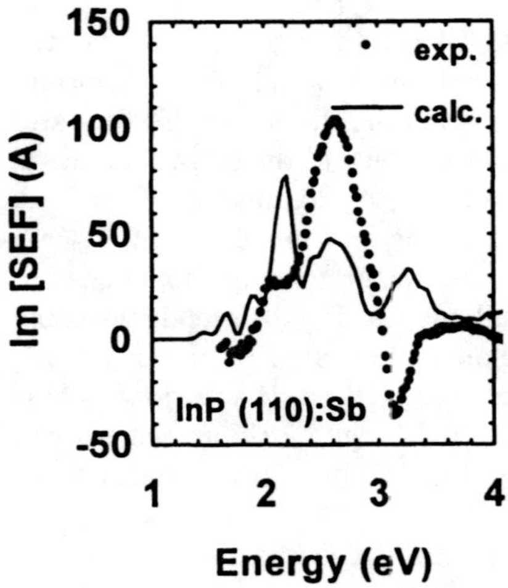

Fig. 3.13. Comparison between the experimental (*dots*) and calculated (*continuous line*) surface excess function of Sb-monolayer-terminated InP(110). The experimental and the calculated data were determined from the differences between the dielectric functions of Sb-covered and clean surfaces [3.141]

Both techniques have been applied to III–V(110) surfaces terminated by an Sb monolayer in order to characterize their dielectric properties and to compare experimental with corresponding theoretical results [3.102, 3.108, 3.136,3.141]. Figure 3.13 shows the surface excess function of an Sb-monolayer-terminated InP(110) surface determined by ellipsometry [3.141]. For comparison, the surface excess function calculated by a tight binding approach is also given. The surface excess function reveals a strong maximum at 2.6 eV and additionally a pronounced shoulder at 2 eV (the negative feature at 3.15 eV arises from the E_1-gap of the InP substrate). These two structures, which coincide with the maxima of the Raman resonance profile in Fig. 3.12, are characteristic of the surface dielectric properties. Also, the RAS spectra shown in Fig. 3.14 reproduce again prominent features at just those photon energies. A maximum at 2.1 eV is followed by a minimum at 2.65 eV. The RAS experiments give additional information about the preferred polarization direction of the light for interaction with the surface electronic states. The RAS data indicate that the 2.1 eV transition is predominantly polarized along the [001]-direction (i.e., perpendicular to the Sb chains) while the 2.65 eV transition is polarized along $[1\bar{1}0]$.

The preferential polarization of the observed surface transitions can be used to discuss possible assignments of the observed spectral features to specific surface transitions, since transition energies and orbital character of the surface states are known from the band-structure calculation discussed in Sect. 3.3.2 (see Figs. 3.4 and 3.5). For this purpose the surface dielectric anisotropy, $\Delta\epsilon_{\mathrm{s}}$, has to be considered, which is related to the RAS signal by [3.147]:

$$\frac{\Delta r}{r} = 4\pi \mathrm{i}\frac{d}{\lambda} \cdot \frac{\Delta\epsilon_{\mathrm{s}}}{\epsilon_{\mathrm{b}} - 1} , \tag{3.13}$$

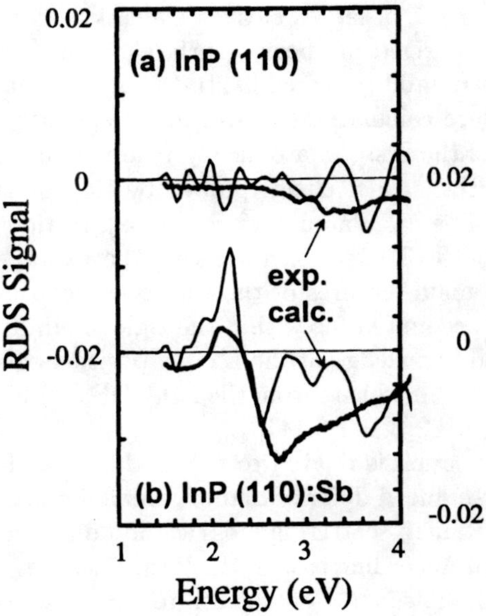

Fig. 3.14. Comparison between the experimental and calculated optical surface anisotropy of clean (**a**) and Sb monolayer-terminated (**b**) InP (110) [3.141]

where ϵ_s and ϵ_b denote the dielectric functions of the surface and the bulk, respectively. Since the surface transitions under consideration occur at photon energies well below the E_1-gap of the bulk material, ϵ_b is, to a good approximation, a positive real quantity. Therefore the real part of $\Delta r/r$ (shown in Fig. 3.14) is related to the imaginary part of $\Delta\epsilon_s$ (peaks in re(Δr) correspond to absorption maxima due to surface transitions).

The 2.1 eV transition being predominantly polarized along the [001]-direction should most likely originate from S_5, S_6 to S_7, S_8 surface transitions. Assuming a preferential polarizability of the covalent bonds along the bond direction, a positive RAS feature is expected from these transitions since the Sb–substrate bonds are oriented along the [001]-direction within the surface plane. Moreover, according to the calculated surface band structure, contributions of these surface states are expected around 2.1 eV since they have approximately this energy separation over large parts of the SBZ [3.59]. Following the polarizability argument, the negative feature at 2.65 eV could be related either to Sb–Sb interchain bonds or to bonds between the first and second substrate layer. The latter would give rise to electronic states distinct from the bulk due to their different chemical environments, i.e., the bonding of the first layer atoms to the Sb monolayer. The calculated band structure reveals that the transitions related to the Sb interchain bonds are at considerably higher photon energy than accessible in the experiment, and thus the modified substrate bonds should account for the 2.65 eV feature. The relevant states may be reliably identified with the A_2-surface-state found in the band-structure calculation. This interpretation is corroborated by the orbital representation of the A_2 state discussed in [3.59].

Apart from the question of whether the observed surface transitions can really be attributed to surface states derived of specific surface bonds (as one would expect within the bond polarizability model [3.2]), the agreement between experiment and band structure calculations proves unambiguously that the surface electronic properties of the first few atomic layers are distinct from those of the bulk, giving rise to the surface effects probed by ellipsometry, RAS and Raman spectroscopy. The calculated surface excess function and surface anisotropy shown in Figs. 3.13,3.14, together with the experimental results, demonstrate that the main features of the surface dielectric properties can be quite satisfactorily accounted for within the tight binding approach. For this purpose, a detailed knowledge of the atomic positions of the monolayer structure (for instance obtained here from the LDA-DFT total energy calculations) is required [3.59, 3.102, 3.108, 3.141].

Since the Raman scattering cross section is related to the modulation of the dielectric function by the phonon-induced deformation of atomic bonds, it should be possible to calculate the Raman scattering cross section directly within the tight binding scheme [3.141]. According to [3.2] the Raman scattering efficiency (in units of [length·solid angle(sr)]$^{-1}$) is quantitatively related to the modulation of the dielectric function $\Delta\epsilon$ by a phonon mode j with amplitude Q_j:

$$S = \frac{\omega_{\mathrm{s}}^3}{(4\pi c_0^2)^2} \frac{\hbar}{2\omega_{\mathrm{ph}}} V_{\mathrm{c}} \left[n_{\mathrm{BE}}(T) + 1 \right] \left| \frac{\Delta\epsilon_{\alpha,\beta}}{\Delta Q_j} \right|^2 , \tag{3.14}$$

where ω_{s} denotes the frequency of the scattered light (which is approximately equal to that of the incident light), ω_{ph} is the phonon frequency, V_{c} the scattering volume, n_{BE} the Bose–Einstein occupation probability, α, β refer to the polarization of incident and scattered light, and $\Delta\epsilon_{\alpha,\beta}/\Delta Q_j$ corresponds to the linear term of (3.8) in Sect. 3.2.2. This expression follows from (3.12) taking the spectrometer throughput and reflection losses at the sample surface into account. In order to determine $\Delta\epsilon_{\alpha,\beta}/\Delta Q_j$ the dielectric function must be calculated for different atomic geometries, i.e., for the ground state and for atomic positions shifted according to the phonon eigenvectors Q_j of the surface vibrational modes [3.141].

Figure 3.15 shows the result obtained for the three intense 1, 2, $3A'$–modes of Sb-terminated InP(110) having a diagonal Raman tensor with two components a, b for polarization along the $[1\bar{1}0]$- and [001]-directions on the (110) surface, respectively [see Sect. 3.3.4 and (3.11)]. The calculated Raman cross section explains the experimental finding of different Raman scattering intensities for parallel light polarization either along $[1\bar{1}0]$ or [001] using the 2.41 eV Ar-laser line for excitation: At 2.41 eV the absolute values of the tensor components a (thick lines) are much higher than those of components b (thin lines) for the $1A'$ and 2A′–modes, in contrast to the $3A'$–mode, where the a and b components have almost equal absolute value. Basically the same behavior is found in the experiments (see Sect. 3.3.4, Fig. 3.8): The $1A'$– and $2A'$–modes are strong under ($[1\bar{1}0]$,$[1\bar{1}0]$) polarization and almost absent

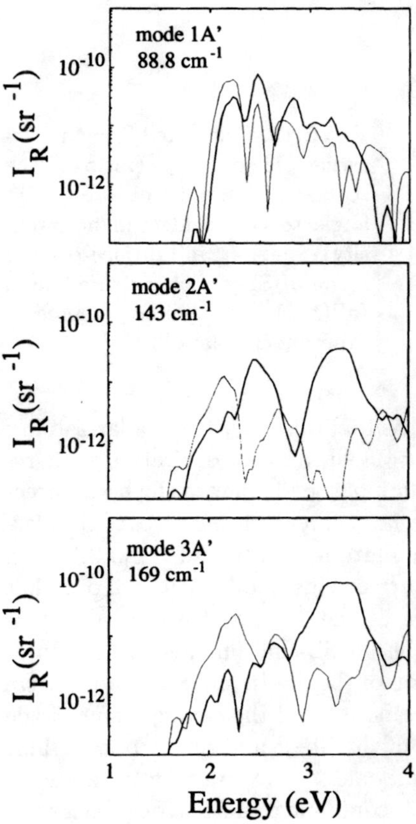

Fig. 3.15. Calculated Raman scattering efficiency I_R (ratio of the scattered intensity per unit solid angle to the incident light intensity for the 1, 2, and $3A'$ surface phonon modes of an Sb monolayer terminated InP(110). The thick (thin) lines correspond to the resonance profiles of the a (b) Raman tensor components [3.141]

under ([001],[001])-polarization, while the $3A'$–mode is observed at almost equal intensity in both configurations. Also, the general spectral line shape of the resonance profiles for the a components of the 1, 2, and $3A'$–modes reproduces quite well the experimentally determined resonance curves for ([1$\bar{1}$0],[1$\bar{1}$0]) polarizations. Both resonance profiles of the 1 and $2A'$–modes show a pronounced maximum around 2.5 eV, whereas that of the 3A′–mode reveals a weak maximum around 2.6 eV followed by an additional increase to higher photon energies.

The result of the corresponding calculation for the $2A''$ surface phonon mode of Sb monolayer terminated InP(110) is shown in Fig. 3.16. The Raman tensor of the A″ modes is non-diagonal, having only one independent component [see Sect. 3.3.4, (3.11)]. The agreement between calculated and experimental resonance profile is not as good as in the previous three cases. However, the main experimental finding of a pronounced resonance at 2 eV is reproduced.

Additional information about the electronic states involved has been derived in [3.141] on the basis of the bond polarizability model. In this model,

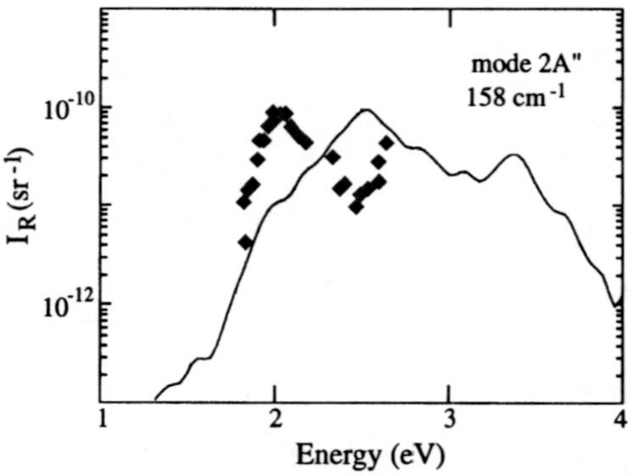

Fig. 3.16. Calculated Raman scattering efficiency I_R (ratio of the scattering intensity per unit solid angle to the incident light intensity) for the $2A''$ surface-phonon mode of Sb-monolayer-terminated InP(110). The symbols represent experimental data [3.141]

the interatomic bonds are treated as individual dipoles whose polarizability under the influence of an external electric field, and its derivative with respect to the bond length, is larger along than perpendicular to the bond direction [3.2]. This assumption is known to be valid at least in the cases of the electronically very similar diamond structure materials C, Si and Ge [3.2, 3.141]. Using the atomic displacements of the surface eigenmodes it turns out that the a-components of the Raman tensors of the 1, 2, and $3A'$–modes contain contributions from modulations of the Sb–Sb-chain bonds and the In–P bonds, but do not contain any contribution of the Sb–In and Sb–P bonds. On the contrary, the b-components of the A'–modes and the c-component of the $2A''$–mode contain also contributions from the Sb–P and Sb–In bonds. This result may explain why the b- and c-components give rise to a pronounced resonance at 2 eV which is absent in the a-component: The strong feature at 2 eV observed in the resonance profile of the A''–mode must be correlated to surface states derived from the Sb–substrate atom bonds S_5, S_6 to S_7, S_8, whereas the 2.5–2.65 eV transition originates from the In–P bonds between the first and second substrate layers. It should be pointed out that this result agrees with the assignment made on the basis of SE and RAS experiments and the LDA-DFT band structure calculations (Figs. 3.4,5,13,14).

In the above discussion concerning the resonance profiles of the surface phonon modes it has been established using Sb monolayers on III–V(110) as an example that the scattering mechanism is based on the deformation potential which causes a dynamic modulation of surface electronic properties by the surface phonon modes. The vibration of atoms in the first few layers determines the vibrational properties of the observed modes as well as the electronic properties displayed by the respective resonance profiles. This mechanism is rather general and should apply to any surface. Thus Raman scattering from surface vibrational modes should be observable also on other, clean as well as adsorbate covered semiconductor surfaces. The remaining differences between experiment and theoretical modelling of the Raman scattering efficiency are most likely due to the simplifications inherent in the calculations, such as neglecting many body effects (e.g., excitonic interaction) which could yield a significant contribution.

The role of the Sb monolayer in the observation of surface resonant Raman scattering is to modify the surface electronic properties so that the electronic resonances of the surface phonons occur at photon energies which are well separated from the bulk interband critical points of the respective III–V substrates. Consequently, large Raman scattering efficiencies occur for the monolayer vibrational modes at photon energies for which the scattering by bulk modes is weak. This helps substantially to separate surface from bulk contributions; in cases where both bulk and surface resonances are close together a strong bulk signal may mask the weaker surface signal. This, in fact, is observed in the case of Sb monolayers on GaP(110), where the surface resonances are very close to the E_0 bulk band gap [3.103].

The absolute value of the Raman scattering cross section for the monolayer modes, however, is not at all an unusually large quantity specifically related to the Sb monolayer systems. This can be easily seen if one compares the scattering cross sections of the monolayer modes under surface resonant conditions to that of the TO phonon bulk modes under bulk resonant conditions. The absolute scattering intensity of the A′-mode for excitation at 2.41 eV is about 20 times smaller than that of the InP TO phonon for laser excitation at 3.05 eV, close to the bulk E_1 band gap (3.16 eV [3.51, 3.145]). This intensity ratio is accounted for by the different scattering volumes of the A'– and the bulk TO–mode: for the InP TO phonon the scattering volume is limited by the light attenuation to around 100Å (half of the light penetration depth of 21 nm at 3.05 eV) near the E_1 gap. A 20 times smaller scattering volume, according to the relative intensities, would correspond to around 5Å, i.e., approximately twice the distance of neighboring atomic (110) layers, as expected due to the confinement of the surface phonon eigenmodes to the first few atomic planes. Consequently, for the surface phonons the resonant cross section is of the same order of magnitude as for deformation potential scattering by III–V bulk phonons, which supports the interpretation that similar scattering mechanisms are operative and no surface specific enhancement mechanisms occur. We thus conclude that Raman scattering from surface phonons is a general feature observable for surfaces of solids having a reasonable deformation potential cross section for bulk phonons.

3.4 Monolayer Terminated Si(111) and InP(100) Surfaces

Besides the Sb monolayers on III–V(110) surfaces which we have discussed as model systems for understanding surface Raman scattering mechanisms, other examples of Raman scattering from surface vibrations can be found in the literature. The first two examples discussed in this section consider again structurally and electronically rather well understood systems: Si(111) surfaces terminated either by an As monolayer or by atomic hydrogen. They represent, because of their simplicity, model systems for Si surfaces, in some sense the counterpart to the Sb monolayers on III–V surfaces. A third example of surface Raman scattering treated in the following will deal with a more complex case, i.e., the S terminated InP(100) surface. More recently, surface modes have also been detected by Raman spectroscopy for Sb/Si(001). First results for that example can be found in [3.152].

3.4.1 Surface Vibrations of Arsenic Terminated Silicon(111)

Usually semiconductor surfaces reconstruct with substantial atomic rearrangements compared to the bulk crystal structure. A well known example of such type is the quite complex (7×7) superstructure observed on clean Si(111) surfaces. Unsaturated dangling bonds at the surface are the main driving force inducing surface reconstruction. Thus, by means of an adequate passivating layer the surface structural properties may be greatly affected. One example of such passivating monolayer systems is As on Si(111) which has attracted a lot of interest for many years [3.153]. By adsorption of an As monolayer it has been shown that the (7×7) reconstruction of clean Si(111) is replaced by a simple p(1×1) surface structure. The restructuring of the surface is driven by the saturation of dangling bonds accompanied with the replacement of group-IV surface atoms by group-V atoms: on ideal surfaces (truncated bulk) each surface Si atom would have one dangling bond orbital half-filled with one electron, which is unfavorable from an energetic viewpoint. Replacing the surface Si by As in the same structural configuration creates an energetically favorable doubly occupied dangling bond orbital per surface As atom. Thus, a simple and quite stable surface passivation is achieved by As monolayer termination. The corresponding structure, where the outermost atomic Si layer is replaced by As, is shown in Fig. 3.17.

The precise locations of the As atoms terminating the Si(111) surface have been experimentally determined by XSW, STM, and also theoretically by ab initio total energy minimization within the LDA-DFT theory [3.154, 3.155, 3.156]. It was found that the As adatoms are located in a near-bulk-like position of the Si lattice, with only a slight outward relaxation of 0.17Å. The electronic surface band structure was also calculated by DFT-LDA including quasiparticle corrections [3.156]. Using the so-called GW correction, the underestimate of electronic excitation energies, a common artifact in normal LDA calculations, is remedied. Occupied and unoccupied surface states were investigated by means of STM spectroscopy, angle resolved photoemission and inverse photoemission spectroscopy [3.154, 3.157, 3.158]. Concerning the electronic structure, excellent agreement between calculations (after including quasiparticle corrections) and experimental results was obtained, Fig. 3.18 shows both the calculated surface band structure and experimental data for the occupied and empty surface states. The orbital character of the surface states has also been discussed [3.156]. The associated charge

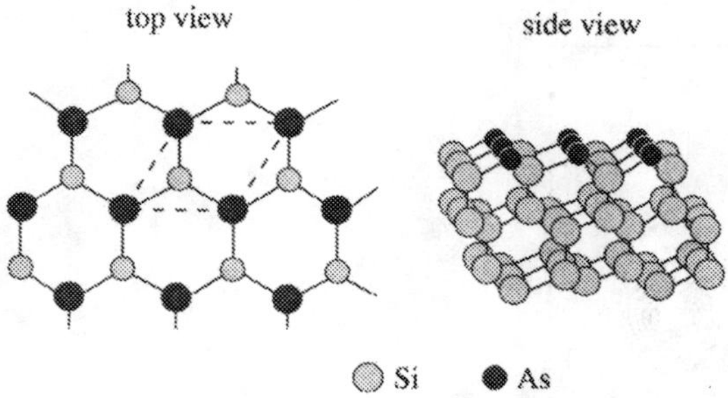

Fig. 3.17. Atomic structure of the As-terminated Si(111) surface. The outermost Si layer is replaced by As atoms [3.153]

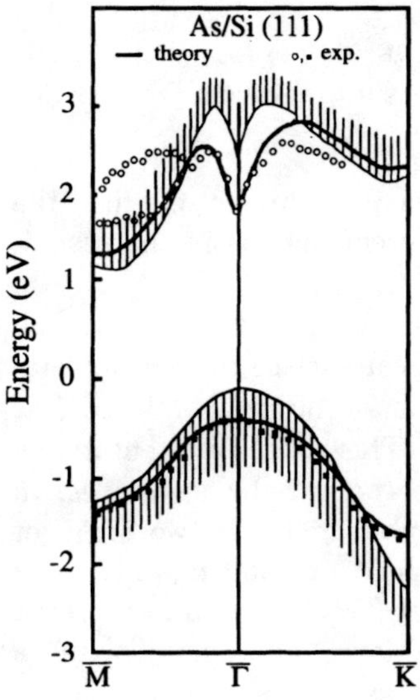

Fig. 3.18. Surface band structure of As/Si(111) calculated in the quasiparticle approach. For comparison, experimental data of the empty surface bands determined by inverse photoemission spectroscopy and occupied surface bands determined by angle-resolved photoemission spectroscopy are included. The solid lines denote calculated surface bands, hatched areas correspond to the projected bulk bands, dots represent experimental data [3.156, 3.158]

density distributions at the $\bar{\Gamma}$ point of the surface Brillouin zone are shown in Fig. 3.19. The occupied state has predominantly p_z character associated with the doubly occupied As dangling bond orbital. However, it extends farther into the bulk due to resonance with bulk electronic states. The empty state has a mixed character of s and p_z atomic orbitals derived from the conduction band of the Si bulk and is confined to the surface layer due to lack of overlap with the bulk bands at $\bar{\Gamma}$. Summarizing, the structural and electronic properties of the As/Si(111) system are well understood.

The surface vibrational properties of the As/Si(111) system have also attracted considerable interest. The acoustic surface modes have been studied by HAS and by theoretical approaches [3.159, 3.160, 3.161]. According to these studies, the surface dynamical properties associated with the macroscopic elastic surface modes can be understood with a mass defect model,

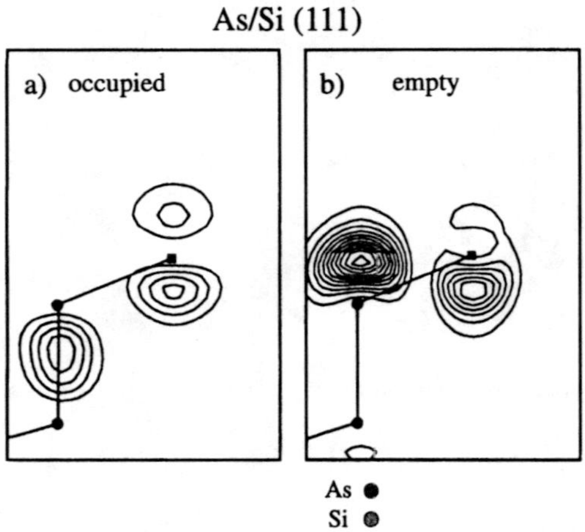

Fig. 3.19. Charge density distribution associated with the **(a)** occupied and **(b)** empty surface states (resonances) at the $\bar{\Gamma}$ point of the surface Brillouin zone [3.156]

by describing the As-terminated surface as an ideal Si surface in which the surface atomic mass has been replaced by a different one, as in the case of isotopic substitution [3.159, 3.160, 3.161].

Our understanding of the optical surface phonons, in contrast, is still incomplete. In a recent HREELS study the optical surface modes were investigated [3.162] and a weak optical mode around $200\,\mathrm{cm}^{-1}$ detected, as previously seen in the HAS experiments [3.159]. This mode was tentatively attributed to a surface resonance coupled with z-polarized bulk modes. In the higher frequency range which is not accessible to HAS, two stronger surface related losses around $356\,\mathrm{cm}^{-1}$ and $520\,\mathrm{cm}^{-1}$, exhibiting only weak dispersion, were found by HREELS. The mode at $356\,\mathrm{cm}^{-1}$ was interpreted as a surface mode localized in the underlying Si layer, i.e., not involving a vibration of the topmost As layer, since HREELS results obtained on H-terminated Si(111) showed a surface mode appearing at approximately the same eigenfrequency. From the experimental result that a large dipole scattering contribution was observed by HREELS for this mode, it was concluded that it should have a z-polarized character, i.e., atomic displacements mainly along the direction normal to the surface [3.162]. The $520\,\mathrm{cm}^{-1}$ mode, coincident in energy with the Si bulk TO phonon, was associated with the Lucas mode, i.e., a shear-horizontal optical surface mode [3.162].

Optical surface phonon modes of the As-monolayer-terminated Si(111) surface have also been observed by Raman spectroscopy. The As-monolayer-terminated samples used in the Raman studies were prepared using Si(111) wafers capped by a thick As layer in an MBE growth vessel. Subsequently, after re-introducing the sample into a separate UHV analysis chamber, the As-monolayer-terminated Si(111) surface was prepared by thermal desorption of the excess As cap [3.163, 3.139]. A strong surface phonon mode was

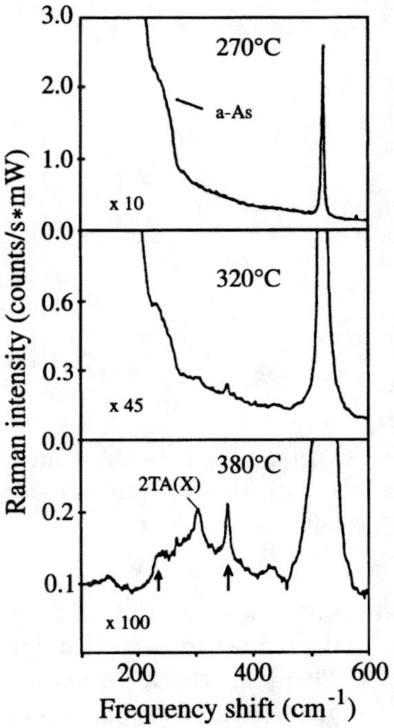

Fig. 3.20. Raman spectra recorded during thermal desorption of an As cap from Si(111). An As-monolayer-terminated surface is obtained after annealing to 380°C. Besides a weak shoulder around 230 cm^{-1} a strong surface mode appears at 356 cm^{-1} (indicated by arrows) [3.163]

identified by Raman scattering at 356 cm^{-1}, in agreement with the HREELS results [3.162], moreover, a weak indication of a lower energy mode around 230 cm^{-1} was reported (Fig. 3.20).

The polarization selection rules and the resonance dependence of the Raman scattering cross section have been investigated for the surface mode at 356 cm^{-1} [3.139]. Figure 3.21 shows Raman spectra under well-defined polarization configurations. The surface mode at 356 cm^{-1} is observed for parallel as well as for crossed polarization configurations. The group-theoretical analysis of the As-terminated Si(111) surface considering a two-atomic base (one As atom in the outermost and one Si atom in the first substrate layer) reveals three optical surface modes. One mode is of A symmetry corresponding to displacements along the C_3 axis (i.e., surface normal, denoted as z direction), the others are a doubly degenerate mode of E symmetry representing in-plane displacements along the x- and y-directions, respectively. The Raman tensors for these modes in the two-dimensional surface coordinate system read

$$\tilde{R}(A_z) = \begin{pmatrix} a & 0 \\ 0 & a \end{pmatrix} , \quad \tilde{R}(E_x) = \begin{pmatrix} 0 & c \\ c & 0 \end{pmatrix} , \quad \tilde{R}(E_y) = \begin{pmatrix} -c & 0 \\ 0 & c \end{pmatrix} . \tag{3.15}$$

Therefore, in the case of a mode displacement along the z-direction (A-symmetry) the Raman signal should appear only under diagonal (x, x) or (y, y) configurations. In the case of displacements in the xy-plane, both diagonal and crossed polarization configurations should give symmetry-allowed

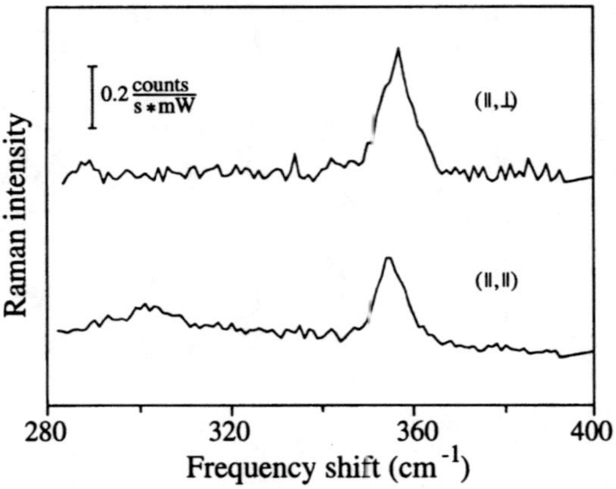

Fig. 3.21. Raman spectra of the surface mode at 356 cm^{-1} on As-terminated Si(111) recorded for parallel and crossed polarizations. In both configurations the surface mode appears with almost equal intensity [3.139]

contributions. The experimentally determined results shown in Fig. 3.21 reveal an E-symmetry for the 356 cm^{-1} mode, i.e., the mode is observed under both parallel and crossed polarization directions. Therefore, the A-symmetry character of the 356 cm^{-1} mode deduced from the HREELS experiments [3.162] is not confirmed. In addition, the E-symmetry of the 356 cm^{-1} mode was recently also obtained by a calculation of the mode eigenvectors using an interplanar force model based on ab initio LDA total energy calculations [3.166].

Figure 3.22 shows the calculated surface dispersion for comparison with the experimental results discussed above. For the acoustic surface branches the new calculation confirms mainly the validity of the earlier Bond Charge Model (BCM) [3.160], but also, the nature of the optical surface branches is clarified. The surface mode appearing at 230 cm^{-1} in the gap between LA and TA bulk branches corresponds to displacements of the As atoms parallel to the surface, while the two underlying Si layers basically move along the surface normal direction. When approaching the $\bar{\Gamma}$ point of the SBZ this mode decays into the mainly z-polarized bulk continuum due to strong coupling to bulk modes. The resulting surface resonance leads to the weak features observed by HAS and HREELS in this frequency regime. The higher frequency mode appearing around 345 cm^{-1} at the zone center of the SBZ is found to be a localized surface mode. This mode corresponds to the 356 cm^{-1} mode found by Raman spectroscopy and HREELS. Remarkably, at the center of the Brillouin zone a purely xy-polarized displacement pattern is found (Fig. 3.23), in contrast to the previous assumption based on the HREELS result, but in agreement with the Raman selection rules. The displacements of the surface As are 1/3 of that of the second layer Si, whereas displacements in the following layers are nearly zero. For larger $\boldsymbol{k}_{||}$, a mixed xyz-displacement

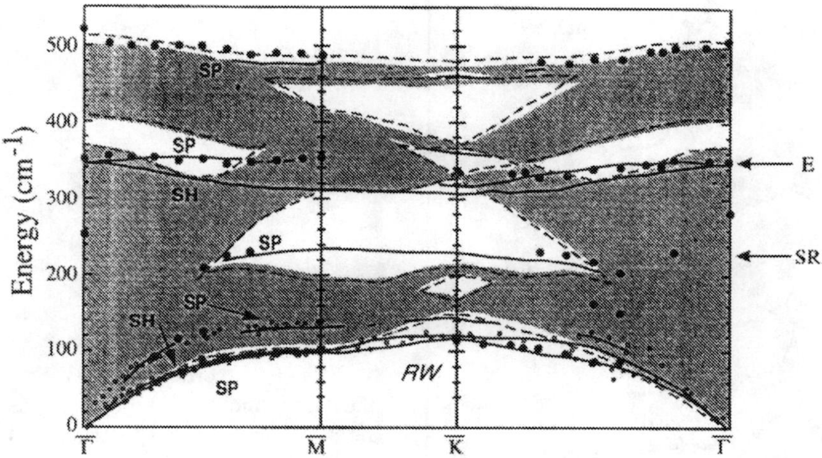

Fig. 3.22. Surface phonon dispersion of As/Si(111) (1×1) [3.166]. The shaded regions correspond to the projected bulk band structure. The thick lines represent calculated modes localized at the surface (RW: Rayleigh Wave, SP: Shear Parallel, SH: Shear Horizontal). Open and full circles represent HAS [3.159] and HREELS [3.162] data, respectively. On the right-hand axis the surface modes determined by Raman spectroscopy are indicated (E: E-symmetry surface mode, SR: Surface Resonance) are indicated

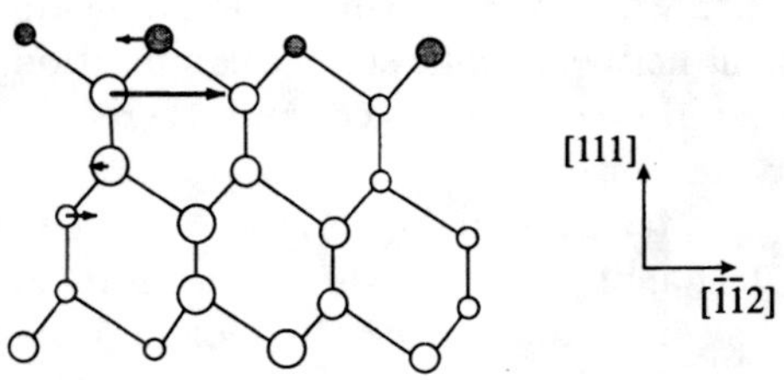

Fig. 3.23. Atomic displacements of the 345 cm^{-1} mode of As/Si(111) at $\bar{\Gamma}$ [3.166]

shows up and the mode splits into a Shear Horizontal (SH) and Shear Parallel (SP) branch. SH denotes an displacement perpendicular to sagittal plane, SP parallel to sagittal plane (sagittal means the plane defined by $\boldsymbol{k}_{||}$ and surface normal). Since in the Raman experiments the $\boldsymbol{k}_{||}$ momentum transfer is nearly zero, they reflect the xy-polarized mode at the 2D zone center.

Another question to be addressed is the resonance behavior. Due to the surface confined nature of the 356 cm^{-1} mode (345 cm^{-1} in the calculations) a contribution of surface electronic transitions to Raman scattering should be expected. Based on the calculations of *Hybertsen* and *Louie* [3.156] surface excitations should occur at transition energies around 3 eV, where the surface bands extend parallel in k-space along the $\bar{\Gamma}$–$\bar{M}$ direction (see Fig. 3.18). Consequently, surface contributions are expected below the E_1, E_0' bulk critical points of Si at 3.3 eV and 3.4 eV, respectively [3.164]. Unlike the Sb/III–

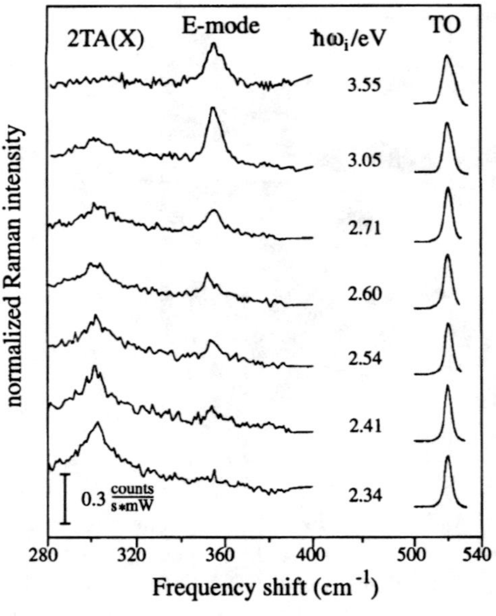

Fig. 3.24. Raman spectra of the 356 cm^{-1} surface mode on As/Si(111) recorded with different laser photon energies $\hbar\omega_i$ as indicated with each spectrum. All spectra are normalized to equal Si TO-phonon height [3.139]. The structure labeled 2TA(X) is a second-order phonon process in Si

V(110) case, the expected contribution of these surface transitions has not yet been clearly confirmed by optical spectroscopy techniques. In preliminary ellipsometry experiments As termination of the Si(111) surface was found to modify the dielectric function mainly in the vicinity of the bulk critical points of Si [3.165]. However, in that experiment neither atomic surface composition nor surface electronic band structure could be cross-checked, and, therefore, the surface conditions were not clearly defined.

Raman resonance experiments on As-terminated Si(111) are shown in Fig. 3.24. The successful As surface termination is evident in the Raman experiments from the appearance of the corresponding surface phonon peak. The spectra, normalized to equal TO phonon intensities, reveal an increase in intensity of the surface mode for incident photon energies approaching 3 eV, while above 3 eV it remains constant. One reason for the change in intensity ratio between surface and bulk contributions is the decrease of the light penetration depth with increasing photon energy [3.164]. Due to this effect a significant increase of the surface contribution between 3 eV and 3.5 eV should occur. This, however, is not observed in the experiment which suggests that surface transitions predicted at 3 eV in the calculations in fact contribute to the Raman scattering efficiency. So far the Raman resonance experiments, however, have been of preliminary nature, because in the range above 3 eV they are based on only two data points (3.05 eV and 3.55 eV excitation).

Although further experiments would be required to allow a detailed interpretation of the resonance of the Raman cross section, the results obtained on As-terminated Si(111) are consistent with the deformation potential scattering involving surface states as found for the Sb-monolayer model system. Furthermore, the usefulness of eigenfrequencies and symmetry selection rules from Raman experiments for the determination of structural and dynamical surface properties is demonstrated once again by this example.

3.4.2 Hydrogen-Terminated Silicon(111)

Usually, atomically clean semiconductor surfaces are prepared under UHV conditions by preparation techniques such as cleavage, sputtering/annealing, or thin layer growth. A remarkable exception is represented by hydrogen-terminated Si surfaces. As has been demonstrated by a number of UHV techniques, well defined atomically flat and clean Si surfaces can also be prepared ex-situ by a wet chemical treatment, in particular for H terminated Si(111) [3.39, 3.167, 3.168]. In brief, the chemical preparation consists of an initial degreasing in organic solvents followed by repeated cycles of chemical oxidation and etching by a Shiraki etch and buffered HF, respectively, and a final etching in ammonium fluoride solution [3.39]. Besides the large interest into wet chemical preparation procedures for technological reasons, a lot of fundamental research has been devoted to the H/Si(111) surface. In contrast to other UHV-based preparation techniques for Si(111) it has been demonstrated that the wet chemical H-passivation represents a unique way to prepare well ordered Si(111) surfaces. Furthermore, these H-passivated Si(111) surfaces are of unreconstructed (1×1) symmetry, close to the ideal surface produced by bulk truncation [3.167, 3.168, 3.170].

H-terminated Si(111) surfaces have been intensively studied using HREELS [3.173, 3.174] and IRAS [3.167, 3.171, 3.172]. Since the H–Si vibrations at the Si surfaces are dipole active local vibrations, they have been successfully monitored by IRAS and HREELS to study the degree of surface perfection. More recently, the vibrational properties of regularly stepped Si(111) surfaces oriented 9° off towards the $[\bar{1}\bar{1}2]$ direction have been investigated by Raman spectroscopy [3.175, 3.175a, 3.175b]. On these miscut surfaces a regular array of six Si atom wide terraces separated by single bi-layer steps are formed (see Fig. 3.25).

Besides the ordinary H–Si bond stretching vibrations on the terraces, three additional vibrational modes are expected to arise due to a vibrational coupling of silicon mono- and dihidrides at the step edges [3.175, 3.175a, 3.175b, 3.176]. These additional modes show a small energy splitting in the order of one or two meV (8–16 cm^{-1}), which is below the energy resolution of HREELS. Thus, Raman spectroscopy, providing a high spectral resolution, should be an adequate experimental technique to resolve the individual lines. In fact, all four non-degenerate modes, the A mode associated with the H–Si-stretching vibration on the terraces and three step vibrations labeled C_1, C_2, C_3, have been observed by Raman spectroscopy (see Fig. 3.26). The step mode C_1 was associated with a Si–H-stretching vibration of the Si atom at the upper step edges, whereas the C_2, C_3 vibrations have been assigned to out-of-phase

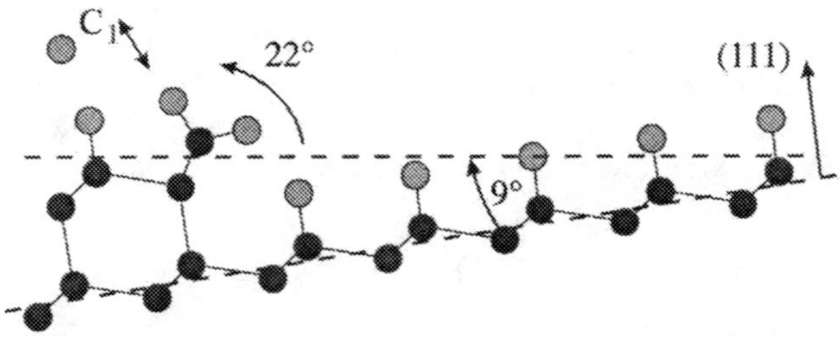

Fig. 3.25. Atomic structure of a H-terminated, nominally (111)-oriented Si surface tilted by 9° off towards [$\bar{1}\bar{1}2$]. A dihydride unit terminates the upper terrace edge. Due to steric effects with the adjacent monohydride unit on the next terrace a strong coupling of monohydride and dihydride vibrations takes place. C_1 denotes a stretching vibration of the upper H bonded to Si atoms at the step edges [3.175, 3.176]

and in-phase combinations of the interacting dihydride and monohydride at the step edges [3.175, 3.175a, 3.175b, 3.176]. The occurrence of Raman signals was shown to depend on the direction of incident and scattered light, in agreement with a description of the H–Si bonds acting as isolated radiating dipoles. A detailed discussion of Raman- and IR-active H–Si vibrations can be found in [3.169]. In the simplified molecular-like picture no scattered light is generated along the line of sight of a H–Si dipole, assuming that the transverse polarizability and its derivative with respect to the bond length are negligible. Thus the C_1-mode cannot be observed if the collected light beam travels along the vibrating bond (configuration D, Fig. 3.26) while it turns up for a beam collected perpendicular to the bond direction (configuration U, Fig. 3.26).

The description of the Raman signal as originating from isolated oscillator dipoles implies that the H-stretching vibrations represent adsorbate vibrations decoupled from bulk properties. This assumption is reasonable from the point of view of mechanical coupling, since the H–Si stretching modes have substantially higher eigenfrequencies than Si phonon excitations. However, an influence of the H–Si bonding on the electronic properties of the first few atomic layers should be expected. In fact, the absolute magnitude of the corresponding Raman scattering efficiency indicates such an electronic modification. For the A-mode on H-terminated on-axis Si(111) surfaces, the Raman scattering efficiency was recently determined for different Ar-laser lines ranging from 457.9 nm to 514.5 nm, i.e., in the blue-green spectral regime [3.177]. In this regime it was found to display an ω^4 spectral dependence, which is expected for non-resonant dipole radiation [see (3.6) in Sect. 3.2.2 and [3.2]]. Consequently, the surface resonances of the H-terminated Si(111) surfaces should occur at considerably higher photon energies, well outside the range accessed in the Raman experiments. This conclusion agrees well with angle resolved photoemission, inverse photoemission spectroscopy and spectroscopic ellipsometry results, showing that possible surface transitions would

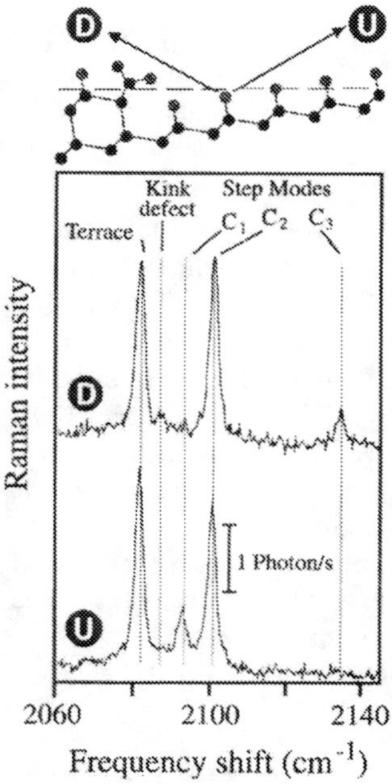

Fig. 3.26. Raman spectra of H-terminated vicinal Si(111) for two different collection directions of the scattered light. D and U denote the respective directions of the collected light beams of the scattered light. The incident beam propagates in the direction perpendicular to D and U. The polarization for the incident beam was chosen approximately along the surface normal [3.175]

occur above 6 eV [3.146, 3.167, 3.168, 3.170]. However, the Raman efficiency for the stretching mode of H-terminated Si(111) shows a strong enhancement of nearly two orders of magnitude in comparison with that for SiH_4 molecules. Thus, the idealized model of undisturbed H–Si-oscillators does not explain quantitatively the Raman signal and, as in the previous examples, microscopic surface properties should play an important role. In agreement with this conclusion ab initio LDA total energy calculations of the atomic structure of the H-terminated Si(111) surface have shown bond length modifications in the few outer layers. The H–Si bond length is enlarged by 0.2Å in comparison to the SiH_4 case of 1.48 Å, the first-to-second layer distance is contracted by 0.3Å and even the second to third is still slightly modified, enlarged by 0.1Å [3.166]. Thus, H adsorption modifies to some extent the electronic properties of the first few Si layers, a fact which may qualitatively explain the difference in the Raman scattering cross section.

3.4.3 Sulfur-Terminated InP(100)

In contrast to Si, on III–V surfaces neither stable native oxide layers usable for surface passivation nor a stable hydrogen termination is formed. Passivation of III–V surfaces, however, can be achieved by sulfur adsorption which therefore is of great technological interest [3.178, 3.179, 3.180]. Although the passivating nature of sulfur layers with respect to chemical and electronic surface properties has been known for some while, the microscopic structure

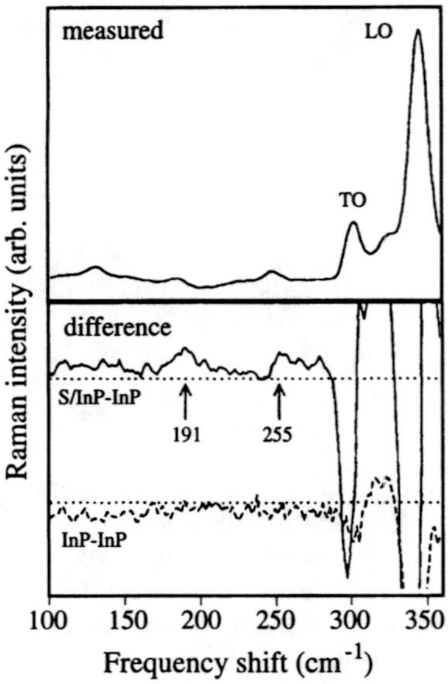

Fig. 3.27. Raman spectra of S-terminated InP(100). The upper panel shows the measured spectra. The bulk-like TO and LO modes appear at $303\,\text{cm}^{-1}$ and $347\,\text{cm}^{-1}$. The lower panel shows difference spectra, upper curve: S/InP(100) minus InP(100), lower curve: InP(100) minus InP(100) reference experiment. The weak peaks at $191\,\text{cm}^{-1}$ and $255\,\text{cm}^{-1}$ have been identified as phonon modes of the S-terminated InP(100) surface, the strong structures represent non-compensated TO- and LO-phonon lines, respectively [3.181]

of sulfur passivated III–V surfaces is still unclear. Based on XPS and LEED studies it was concluded that the InP surface should be covered with a monolayer of S adsorbed at bridge sites on top of a complete second layer of In, forming a (1×1) structure [3.179]. More recently, however, STM experiments resolved a local (2×2) structure producing a (1×1) LEED pattern due to the lack of long range order [3.181]. This discussion has triggered an investigation of the atomic structure of S-passivated InP(100) by ab initio total-energy LDA calculations and by Raman scattering from surface phonons. The atomic structure and the corresponding interatomic forces were calculated in order to compare the calculated surface phonon modes with the experimental Raman results. In fact, weak Raman signals, attributed to microscopic surface vibrational modes, were detected by taking difference spectra of clean and S-terminated InP(100) samples (Fig. 3.27).

In order to derive the atomic structure of minimum energy several different adsorption geometries were considered in the total energy calculations. As a result a (2×2) reconstructed geometry was found, containing two inequivalent S dimers in the outermost layer adsorbed on a complete second layer of indium followed by a third layer of P located in two inequivalent sub-lattice sites. This structure was checked by comparing the experimental mode frequencies with phonon calculations based on interplanar force constants obtained in the LDA calculations. Two surface modes at $190\,\text{cm}^{-1}$ and $257\,\text{cm}^{-1}$ were found to be in excellent agreement with the Raman results showing modes at $191\,\text{cm}^{-1}$ and $255\,\text{cm}^{-1}$. Since the eigenvectors also determined in the calculations involve large displacements of the outermost S and third layer P atoms, the occurrence of these modes has been suggested to be a sensitive and reliable test of the proposed structure model [3.181]. However, it should be mentioned that close to the proposed S-related modes peaks due to second order phonon scattering in InP arise [3.127, 3.128], at $189\,\text{cm}^{-1}$ attributed to 2TA(W) and 2TA(K), and at $252\,\text{cm}^{-1}$ attributed to TO(X)–TA(X) (see Sect. 3.3.4). Moreover, the $189\,\text{cm}^{-1}$ mode has resonant character to the bulk LA(X) mode, as shown in Sect. 3.3.3. Therefore, the assignment of weak features in the difference spectra to surface phonon modes is not unambiguous. However, the correlation of these modes to the surface modification and the coincidence with the calculated mode frequencies supports the interpretation as surface modes. In this case, the $189\,\text{cm}^{-1}$ mode should most likely be a surface resonance, while the $251\,\text{cm}^{-1}$ peak would correspond to a localized surface mode.

3.5 Clean InP(110) Surfaces

The detection of surface phonons on clean semiconductor surfaces by Raman spectroscopy is more difficult than that of phonons at monolayer terminated surfaces. In the examples treated in the previous sections the surfaces studied by Raman spectroscopy were heteroepitaxially terminated, i.e., the surface composition was different from that of the bulk (e.g., Sb/III–V, As/Si, H/Si, S/InP). These cases are favorable for two reasons: Firstly, because of the distinct atomic masses the lattice dynamical properties of the surface differ significantly from those of the bulk. Thus, the eigenfrequencies of surface vibrations should be well separated from those of the bulk. Secondly, the surface electronic band structure is strongly affected by the hetero-termination, and surface electronic transitions (corresponding to critical points in the electronic band structure) should arise at photon energies well separated from bulk transitions.

The second aspect is of particular importance for the Raman detection of surface phonons, since the bulk phonon signal can become unimportant if surface resonant conditions correlate with off-resonance conditions for bulk states. This explains why, for instance, Sb-monolayer vibrations on InP(110) are comparably easy to detect by Raman spectroscopy. However, the results summarized in the preceeding sections have demonstrated that there is no intrinsic limitation of Raman spectroscopy which would prevent the observation of surface phonons on clean surfaces.

In fact, surface phonons have been detected recently by Raman spectroscopy on clean InP(110) [3.8, 3.197, 3.198]. Like the Sb-monolayer terminated III–V(110), the structural and electronic properties of clean III–V(110) surfaces are also well understood [3.121, 3.200]. Considering the surface dynamics, in particular InP has attracted much interest. Due to the large mass difference between In and P a wide gap between acoustic and optical phonon branches opens up which favors the existence of localized surface modes. The studies of the vibrational properties of clean InP(110) surfaces include HREELS [3.201] as well as various calculations using tight binding molecular dynamics simulations [3.202], density-functional perturbation theory (DFPT) [3.203], the phenomenological bond-charge model (BCM) [3.204], and the DFT-LDA frozen-phonon method [3.205]. The atomic structure is similar to the Sb monolayer terminated surface (see Fig. 3.3 in Sect. 3.3.2) except for the relaxation of the clean surface. The surface geometry of InP(110) is common for III–V(110) compounds. Relaxation of the InP surface leads to a tilting of the group-V element outwards and of the group-III element inwards with a buckling angle of 27.1–31.1° [3.203, 3.206]. The surface unit cell has a mirror plane parallel to the [001]-direction (point group C_s). Therefore, the same symmetry arguments hold for the clean InP(110) surface as for the Sb monolayer terminated one and at $\bar{\Gamma}$ the surface modes can be classified as even (A') or odd (A'') with respect to the mirror plane.

As in the Sb-monolayer terminated case, three A'' and six A' modes exist if a four-atom base (first and second InP layer, one In and one P atom per layer) is chosen. BCM calculations [3.204] predicted the eigenenergies of the A'–modes at $\bar{\Gamma}$ to be 9.1 meV (73.4 cm^{-1}) and 11.49 meV (93 cm^{-1}) (overlapping with the acoustic branches), to 28.13 meV (226 cm^{-1}) and 34.77 meV (280 cm^{-1}) (in the gap between acoustical and optical branches), to 37.09 meV (299 cm^{-1}) (overlapping with optical branches) and to 43.3 meV (349 cm^{-1}). The DFPT calculations reveal basically very similar results. At the Brillouin zone center modes at 9 meV (72.6 cm^{-1}) and 14 meV (113 cm^{-1}) overlapping with the acoustical branches are found, the two gap modes appear at 32 meV (258 cm^{-1}) (lower gap mode) and 34 meV (274 cm^{-1}) (upper gap mode), less much separated than in the BCM calculations, and a surface mode shows up above the bulk optical branches at 43.6 meV (352 cm^{-1}), close to the upper border of the bulk continuum at $\bar{\Gamma}$. The high-energy surface mode was consistently interpreted as related to the Fuchs–Kliewer (FK) phonon [3.203, 3.204, 3.205, 3.207]. Experimentally, using HREELS, surface phonon modes were observed at 9.5 meV (77 cm^{-1}), 33 meV (266 cm^{-1}) and 42.3 meV (341 cm^{-1}) (FK phonon). However, due to the low energy resolution of HREELS, the two gap modes could not be resolved and, instead, appeared as one mode around 33 meV [3.201]. Additionally, a surface phonon mode at 19.6 meV (158 cm^{-1}) was found.

A common result of the calculations is that two localized surface modes of A' symmetry within the band gap between the acoustical and optical bulk-phonon branches (28–36 meV at the $\overline{\Gamma}$ point) are predicted. The displacement patterns of these two modes obtained from the DFPT calculations are shown in Fig. 3.28. Since these modes do not overlap with bulk-phonon modes (see Fig. 3.33), the atomic displacements are confined to the first two atomic layers, i.e., they are microscopic surface modes. An experimental proof of these two theoretically predicted gap modes, not resolved in the HREELS data should be possible by Raman spectroscopy due to its far superior resolution, should be able to resolve. Spectra for the clean surface of (n-doped)

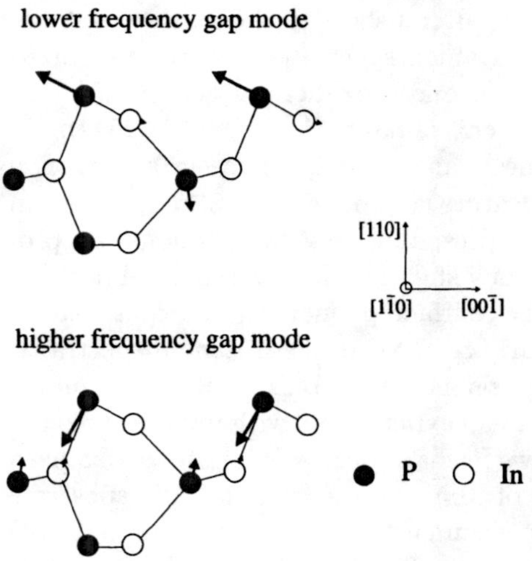

Fig. 3.28. Atomic displacements of the lower and higher surface gap modes at the InP(110)-surface according to DFPT calculations [3.203]

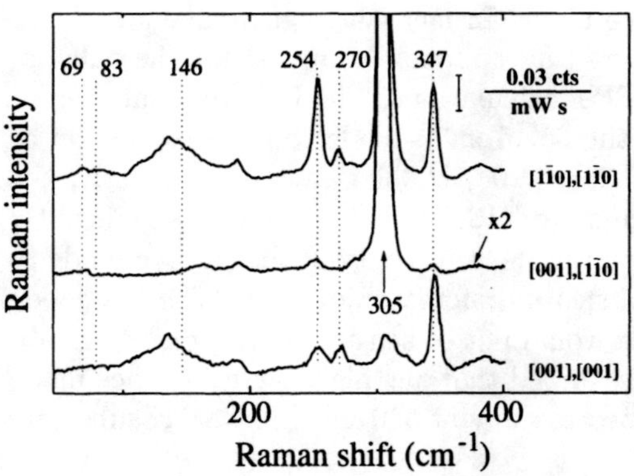

Fig. 3.29. Raman spectra of the clean InP(110) surface for crossed ([001],[1$\bar{1}$0]) and parallel ([1$\bar{1}$0],[1$\bar{1}$0] and [001],[001]) polarization configurations obtained with a laser photon energy of 3 eV. The instrumental resolution is 7 cm^{-1} [3.8]

InP(110) for parallel and crossed polarization configurations of the incident and the scattered light are shown in Fig. 3.29 [3.8]. According to the selection rules obtained from the Raman tensors [3.2], A'–modes should be observed in parallel and A''–modes in crossed polarization configuration only. For these polarization configurations several peaks are observed. The most significant one, at 305 cm^{-1}, is the bulk TO-phonon mode, which is symmetry forbidden in the ([001],[001]) configuration, and symmetry allowed otherwise. Surface phonon contributions (marked with dotted lines) are not clearly evident in the raw Raman spectra. The other features not marked with dotted lines are assigned to multiphonon processes [3.127, 3.128, 3.129]. In order to separate surface-from bulk-phonon modes two sets of Raman spectra were taken, one of a clean and one of an oxidized sample after a short-time exposure to atmospheric conditions. Figure 3.30 shows Raman spectra of clean and oxidized samples for the ($[1\bar{1}0],[1\bar{1}0]$) configuration and the difference spectrum (clean minus oxidized). The structures at $69\,\mathrm{cm}^{-1}$, $83\,\mathrm{cm}^{-1}$, $146\,\mathrm{cm}^{-1}$, $254\,\mathrm{cm}^{-1}$, and $270\,\mathrm{cm}^{-1}$, marked by dotted lines, can be assigned to surface phonon modes of A' symmetry. The weak feature at about $251\,\mathrm{cm}^{-1}$ observed in the spectrum of the oxidized sample corresponds to a two-phonon bulk process [3.129,3.127]. Around $350\,\mathrm{cm}^{-1}$, only slightly above the spectral position of the bulk LO phonon mode, another surface phonon related contribution may arise which, however, is not clearly resolved in these Raman spectra.

In order to resolve the high-energy phonon at $350\,\mathrm{cm}^{-1}$ into the assumed surface and the bulk modes [3.197] Raman experiments with improved resolution were performed. Figure 3.31 shows Raman spectra for clean and oxidized InP(110) surfaces with a resolution of $3\,\mathrm{cm}^{-1}$, recorded at 140K substrate temperature. In the spectrum of the clean surface two modes, the bulk LO mode at $346.5\,\mathrm{cm}^{-1}$ and another mode at $352.5\,\mathrm{cm}^{-1}$ are well separated. Moreover, the $352.5\,\mathrm{cm}^{-1}$ mode disappears completely after oxidation, as expected for a microscopic surface mode. In fact, the eigenfrequency of this mode matches very well that of the surface phonon branch above the bulk optical branches obtained in the DFPT-calculations. The Raman results prove that the previous assignment of the $352.5\,\mathrm{cm}^{-1}$-mode to the FK phonon as suggested in the theory work [3.203, 3.204, 3.205] is incorrect: Firstly, the Fuchs–Kliewer (FK) mode is found by HREELS at $342\,\mathrm{cm}^{-1}$, significantly below the Raman mode, and secondly, the sensitivity of the surface mode to the atomic surface structure confirms its microscopic origin. The FK mode being a macroscopic surface mode would not so sensitively respond on modification of the surface. It should be noted that the high-energy surface mode can also be resolved by Raman spectroscopy at $350\,\mathrm{cm}^{-1}$ for 300 K substrate temperature.

In other words, except for the FK mode, all zone center modes detected by HREELS are also found by Raman spectroscopy. By RS, however, the two gap modes and the microscopic surface mode at $352.5\,\mathrm{cm}^{-1}$ which cannot be identified by HREELS are clearly resolved. Figure 3.32 shows for comparison Raman and HREELS spectra in the spectral interval from $200\,\mathrm{cm}^{-1}$ to $400\,\mathrm{cm}^{-1}$. In this range the two gap modes, the high-energy surface mode and the FK mode occur. The Raman spectra do not show the FK phonon which is the most prominent structure in the HREELS spectra. This finding does not mean that the FK mode is Raman-inactive, but it possesses an extremely small scattering cross section for surface polaritons in backscattering geometry. In forward scattering geometry, to the contrary, FK modes are observed in Raman spectra [3.208]. The great advantage of the higher spectral resolution in Raman spectroscopy is illustrated by the comparison. The shaded region in the Raman spectrum marks the spectral range analyzed in the high-resolution experiment displayed in Fig. 3.31.

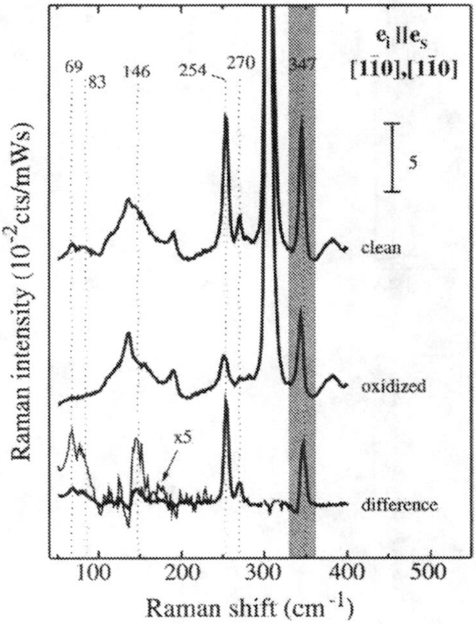

Fig. 3.30. Raman spectra of the clean (top) and oxidized (middle) InP(110) surface and the difference spectrum (bottom) for the parallel ([1$\bar{1}$0],[1$\bar{1}$0]) polarization configuration [3.8]. The spectra were recorded at room temperature with a photon energy of 3 eV. The spectral resolution is approximately 7 cm^{-1}. The surface related structures are marked by dashed lines. The shaded area marks the spectral region selected for the high-resolution Raman spectrum shown in Fig. 3.31

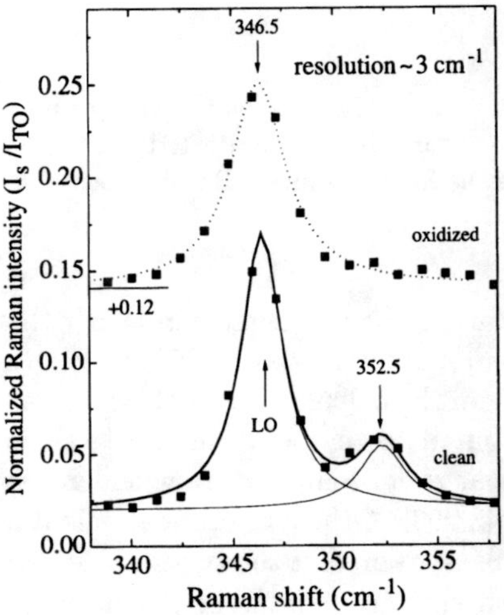

Fig. 3.31. High-resolution Raman spectra of the oxidized (top) and clean (bottom) InP(110) surface in the spectral interval around the LO phonon. The spectra were recorded at 140 K with a laser photon energy of 2.71 eV. The instrumental resolution is 3 cm^{-1}. The surface-related mode at 352.5 cm^{-1} is marked by an arrow. The lines represent Lorentzian fits to the data points [3.197]

Altogether, Raman spectroscopy identifies surface modes with A' symmetry (no A'' modes) at the following eigenfrequencies: 69 cm^{-1}, 83 cm^{-1}, 146 cm^{-1}, 254 cm^{-1}, 270 cm^{-1} and 350 cm^{-1}. Figure 3.33 shows these eigenfrequencies together with surface phonon dispersion data obtained from HREELS [3.201] and from DFPT calculations [3.203]. The eigenfrequencies of the two localized surface phonon modes in the gap between acoustic and

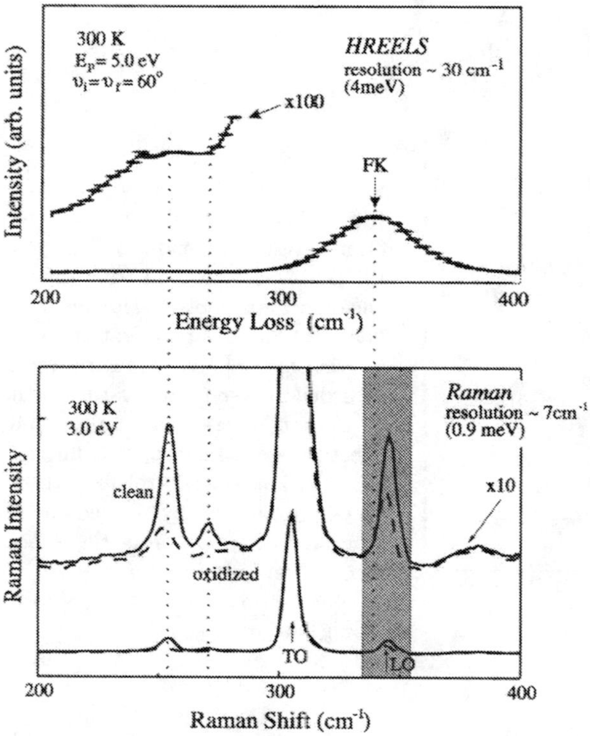

Fig. 3.32. HREELS [3.201] and Raman spectra of InP(110) in the frequency range of $200\,\mathrm{cm}^{-1}$ to $400\,\mathrm{cm}^{-1}$. Surface-phonon modes are marked by dotted lines. The shaded area represents the spectral interval of the high-resolution Raman spectrum shown in Fig. 3.31

optical bulk phonon branches are in remarkably good agreement with the DFPT calculations which predicted the two surface modes at $258\,\mathrm{cm}^{-1}$ and $273.5\,\mathrm{cm}^{-1}$. The difference of $4\,\mathrm{cm}^{-1}$ between the calculated eigenfrequencies and the Raman data can simply be attributed to temperature effects since the Raman measurements were performed at room temperature whereas the calculated phonon energies apply at 0 K [3.203]. Other theoretical calculations, like the above cited BCM calculations, also predict the two surface gap modes in this energy range, however with significantly larger deviation in eigenfrequencies to the experimental results [3.204, 3.205]. Also the localized surface mode at $350\,\mathrm{cm}^{-1}$ ($352.5\,\mathrm{cm}^{-1}$ at 140 K), $5\,\mathrm{cm}^{-1}$ above the respective bulk LO mode at the $\overline{\Gamma}$ point, is accurately predicted by the DFPT calculations at $353\,\mathrm{cm}^{-1}$ [3.203].

The weak feature at $69\,\mathrm{cm}^{-1}$ (Fig. 3.30) corresponds to an A' surface-phonon mode at 73–$77\,\mathrm{cm}^{-1}$ in the DFPT and BCM calculations [3.203, 3.204], also found by HREELS [3.201]. The weak but reproducible structure at about $83\,\mathrm{cm}^{-1}$ (Fig. 3.30) is most likely due to a surface resonance since

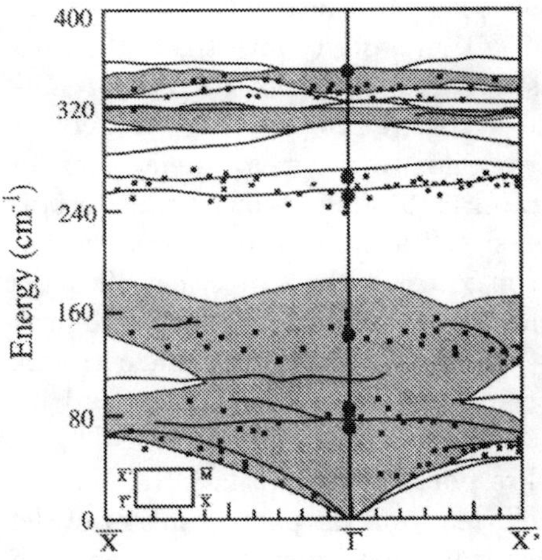

Fig. 3.33. Surface phonon dispersion of InP(110) calculated by DFPT [3.203] (*solid lines*), experimental results of HREELS [3.201] (*solid squares*) and Raman results (*circles*). The shaded areas represent the projected bulk-phonon band structure

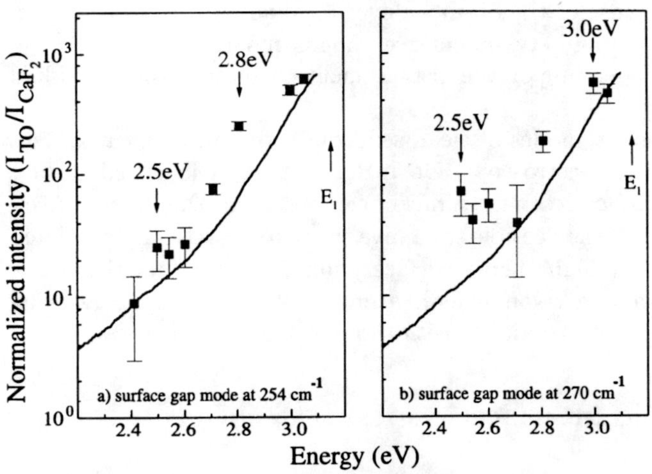

Fig. 3.34. Raman scattering efficiency obtained for the two gap modes (squares, error bars indicated) and the bulk TO phonon (*full line*) of InP(110). The scattering efficiency of the surface modes is distinct: the lower-gap mode shows two maxima around 2.5 eV and 2.8 eV (**a**), that of the higher-gap mode around 2.5 eV and 3 eV (**b**)

in this energy range, at $93\,\mathrm{cm}^{-1}$, an A'-mode is predicted by the DFPT calculations at the $\overline{X}'$ point of the SBZ. Similarly, the weak structure at $146\,\mathrm{cm}^{-1}$ also seen in the HREELS experiments can be assigned to a surface resonance. The DFPT calculations predict a surface phonon mode in this energy range close to the $\overline{X}$ point, but the dispersion branch could not be identified at the $\overline{\Gamma}$ point due to strong coupling to bulk modes.

Additional Raman measurements have been performed with different laser photon energies between 2.41 eV and 3.05 eV in order to investigate the electronic resonance of the surface phonons. Figure 3.34 shows the Raman scattering efficiency obtained for the two gap modes (at $254\,\mathrm{cm}^{-1}$ and $270\,\mathrm{cm}^{-1}$, respectively). The lower gap mode corresponds to the most intense Raman line related to surface phonons on clean InP(110). The resonance of the Raman efficiency of the gap modes shows significant deviations from the behavior of the InP TO bulk mode: two maxima in the scattering efficiency of the lower-gap mode, around 2.5 and 2.8 eV, appear, while the higher-gap mode shows maxima around 2.5 and 3 eV. These maxima correlate with surface transitions which have been found by photoemission and inverse photoemission spectroscopy [3.210], band structure calculations [3.211] and investigations of the surface optical response [3.108,3.212,3.213]. Consequently, the dependence of the Raman efficiency on the photon energy of the exiting laser beam indicates that surface transitions play an important role in the resonance enhancement of the Raman scattering by surface phonons. Moreover, like in the case of the Sb-monolayer-terminated surface discussed in Sect. 3.3.5, the coupling between surface-phonon modes and surface electronic states differs depending on the displacement patterns of the individual phonon mode [3.198].

Summarizing, phonon modes of clean semiconductor surfaces can in fact be detected by Raman spectroscopy. On InP(110) several localized surface modes and several surface resonances have been observed. Due to the excellent energy resolution it was possible to resolve the two theoretically predicted gap modes and another high energy surface phonon for the first time. The surface-gap modes show a resonant enhancement of the Raman scattering cross section related to electronic transitions involving surface states.

3.6 Microscopic Interface Modes

Two different types of interface modes, known as macroscopic and microscopic modes, are found to exist in heterostructures from numerous studies of semiconductor superlattices. The macroscopic modes are accompanied by a macroscopic electric field, analogous to the FK–phonon modes observed on surfaces, thus they also may be referred to as electrostatic interface modes [3.3]. Their eigenfrequencies fall into the range where the real parts of the dielectric functions of the superlattice materials have opposite sign; they follow basically from the electrostatic boundary conditions at the superlattice interfaces [3.3,3.182,3.183,3.184,3.185,3.186,3.187]. The electric fields related to the atomic displacements of the macroscopic modes extend throughout the whole stack of layers, leading to a coupling of the SL layers. Therefore these modes are, strictly speaking, not confined to the interface and should be regarded as a particular type of bulk modes induced by the presence of the interfaces. The second type of interface modes, the so-called

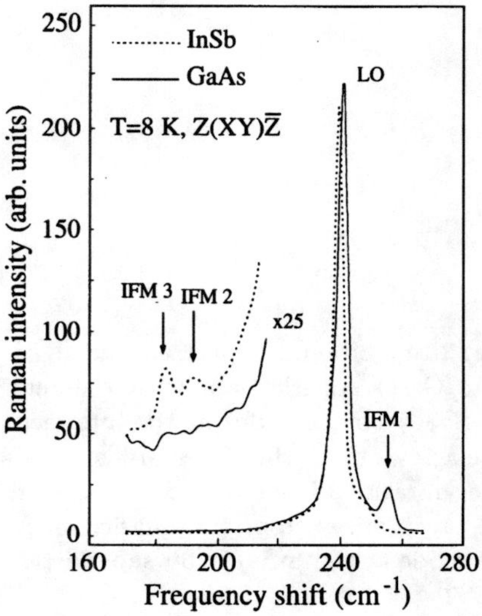

Fig. 3.35. Raman spectra taken from InAs/GaSb superlattices with either In–Sb- or Ga–As-like interfaces. IFM 1 occurs on GaAs-, IFM 2 and IFM 3 on InSb-like interfaces. The strong peak labeled LO arises from the quasiconfined LO mode of the InAs/GaSb superlattice due to the overlapping of the InAs and GaSb LO branches [3.195]

microscopic interface modes, are governed by the mechanical boundary conditions at the superlattice interfaces [3.188, 3.189]. Microscopic interface modes turn up especially for A–B/C–D superlattices where no common anions or cations are present in either material. At these interfaces, of A–D or B–C type, localized modes may exist with eigenfrequencies either in the gap between acoustic and optical bulk phonons or above the optical branches of either superlattice material. Therefore, their extension into the layers is small and they are confined to the interfaces on a microscopic scale. These type of modes is thus comparable to surface phonon modes, while the electrostatic interface modes are specific to superlattices.

Raman studies of microscopic interface modes have been reported for InAs/GaSb, InAs/GaInSb, and InAs/AlSb superlattices [3.188, 3.189, 3.190, 3.191, 3.192, 3.193, 3.194]. For InAs/- GaSb-superlattices, two distinct types of interfaces may be fabricated depending on the growth procedure, either built up of heavy In–Sb- or of light Ga–As-type. The corresponding interface modes are a strongly localized GaAs-type mode at 250 cm^{-1} (IFM 1), above the respective InAs and GaSb phonon branches, and a weakly localized In–Sb mode at 180 cm^{-1} (IFM 3) coupled with acoustic phonon modes in the SL layers [3.188]. Raman spectra obtained from high-quality InAs/GaSb superlattices with either only Ga–As-like or In–Sb-like interfaces are shown in Fig. 3.35. They confirm the existence of the expected interface modes. For In–Sb-like interfaces an additional mode at 194 cm^{-1} (IFM 2) of still unclear origin was observed [3.195].

The localized Ga–As-like mode (IFM 1) has been shown to be an especially sensitive measure of interface quality [3.190, 3.191] and stoichiometry [3.192]. To show this, InAs/GaSb superlattices were grown with controlled

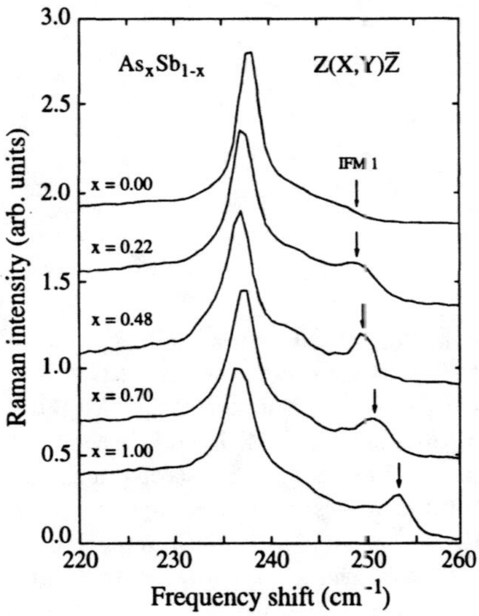

Fig. 3.36. Raman spectra taken on InAs/GaSb superlattices with different As_xSb_{1-x} compositions at the interface. The arrows mark the (Ga–As)-like interface vibration. The strong mode around $237\,cm^{-1}$ arises from the quasiconfined LO mode of the InAs/GaSb superlattice [3.192]

variation of As/Sb stoichiometry at the interface by Migration Enhanced Epitaxy (MEE). The corresponding Raman spectra shown in Fig. 3.36 reveal a systematic dependence of intensity and eigenfrequency for the Ga–As-like interface mode with increasing As concentration at the interface.

For lattice-matched $InAs/Ga_{0.8}In_{0.2}Sb$ superlattices the Ga–As interface mode has been utilized as a direct measure of the dependence of interface quality on growth sequence [3.190]. The corresponding Raman spectra are shown in Fig. 3.37. Beside the strong LO and TO modes of the SL, again the three interface modes IFM 1, IFM 2, and IFM 3 are identified. In the upper spectrum the Sb shutter was kept open during a 10 s growth interruption while changing from the InAs to the GaInSb growth, whereas in the lower two spectra all shutters were closed. The decrease of the localized Ga–As interface mode IFM1, observed in the case of open Sb shutter, are indicative of interfacial mixing due to Sb incorporation into the InAs layer which results in a reduction of Ga–As-type interface bonds. The weakly localized modes IFM2 and IFM3, on the contrary, are not sensitive to the interface conditions.

More recently, the symmetry selection rules of the localized Ga–As-like interface modes were investigated on superlattices exhibiting only Ga–As-type interfaces with Ga–As-bond orientation along one direction (every second interface being of In–Sb-type). For this case it was shown that the Raman scattering can be explained in the deformation potential picture within the bond polarizability model. Strong Raman signals were recorded for polarizations of both the incident and scattered light perpendicular to the interface bonds, i.e., parallel to Ga–Sb and In–As bonds in adjacent atomic layers, whereas only small signals were found for polarizations along the interface bonds (see Fig. 3.38) [3.189]. This result can be understood by assuming much

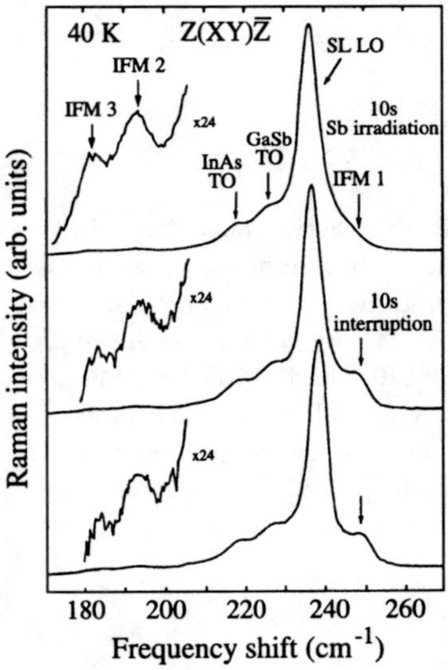

Fig. 3.37. Raman spectra of three different InAs/$Ga_{0.8}In_{0.2}Sb$ samples. The two upper spectra were taken on samples grown on GaAs substrates, the lower one on a sample grown on a GaSb substrate. The top spectrum corresponds to a sample grown with a permanently open Sb shutter. The lower two spectra were taken on samples with the Sb shutter closed during growth interruption while changing from InAs to $Ga_{0.8}In_{0.2}Sb$ growth. A change in the intensity of the localised Ga–As-type interface mode denoted as IFM 1 is clearly observed, while the weakly localised modes IFM 2 and IFM 3 are not significantly affected [3.190]. In the frequency interval between 210 cm^{-1} and 440 cm^{-1} the confined TO modes of the InAs and GaSb layers and the quasiconfined LO mode of the superlattice show up

larger bond polarizabilities of the In–As and Ga–Sb bonds than of the Ga–As bonds, which would be expected at least for photon energies well below the E_1 gap of GaAs (around 3 eV), a condition which prevailed in the experiments. The off-resonance selection rules are therefore in agreement with the expectation from deformation potential scattering, like in the case of the surface modes described in the previous sections.

However, by tuning the laser excitation into resonance with the E_1 gap of GaSb (at 2.05 eV) a strong Raman intensity was found under Y', Y' polarization. This has been interpreted as due to additional Fröhlich scattering contributions [3.196]. In a two-dimensional case, such as the localized surface phonons described in the previous sections, a Fröhlich contribution is not expected (see Sect. 3.2.2). A microscopic calculation of the Raman cross section, similar to that performed for the Sb monolayer system on III–V(110), should resolve this question.

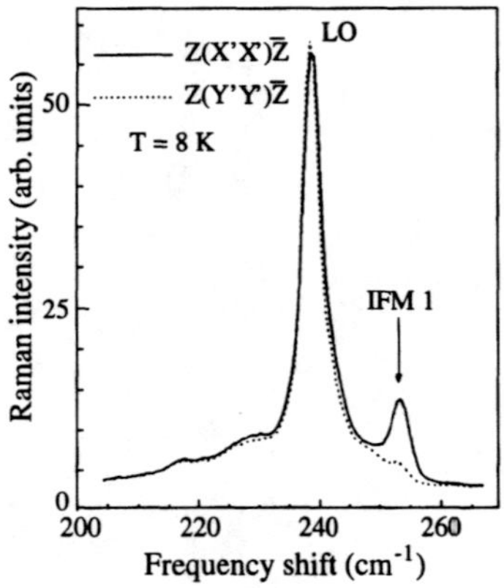

Fig. 3.38. Raman spectra taken from InAs/-GaSb superlattices containing only Ga–As-type interfaces. Two different parallel polarization configurations, $X'X'$ and $Y'Y'$, are shown. X', Y' denote the [110]- and [1$\bar{1}$0]-directions, respectively. The Ga–As-like interface mode occurs only for light polarization perpendicular to the interface bonds [3.189]

3.7 Summary and Conclusions

In this chapter we have reviewed the investigation of phonons localized at surfaces and interfaces by means of Surface Resonant Raman Scattering, a rather novel application of Raman spectroscopy. Only recently, with the improvement of Raman spectroscopic equipment and in particular sensitive, low noise multichannel light detection systems, has Raman spectroscopy of surface phonons become possible. Raman Spectroscopy is now becoming established, alongside HREELS and HAS, as a third experimental technique for the study of surface phonons. In general, Raman scattering has the drawback that in first-order scattering only phonons at the Brillouin zone center can be detected. However, it offers an unparalleled high-energy resolution which even at the present stage has not been fully exploited. In a way similar to Raman scattering and neutron scattering from bulk phonons, surface resonant Raman scattering complements the established HREELS and HAS techniques enabling high-resolution studies of optical surface phonons to be carried out. Most of the presently published studies have been discussed in this article.

Raman scattering is well suited to yield linewidth and frequencies of optical surface phonons with high precision. Their analysis gives access to atomic surface composition and structural information. Thus, processes at surfaces involving changes in atomic composition and structure may be monitored by Raman spectroscopy. Other aspects where Raman spectroscopy could become a powerful tool, but which do not seem to have been touched yet, include the role of anharmonic effects on surface phonons with respect to either temperature or strain.

By taking advantage of electronic resonances, Raman scattering by surface phonons offers an additional information channel. Since the intensity of Raman scattering reflects the interaction of the photons as well as the phonons with surface electronic states, it contains information on the combined density of surface electronic states (critical points) as well as on the phonon deformation potentials of the specific modes. Complementary knowledge on surface states may thereby be obtained.

We hope that this contribution will stimulate other research groups who have access to modern Raman-scattering equipment and ultrahigh-vacuum preparation techniques to use this new technique in the pursuit of surface science. Even in spite of the fact that up to now only a rather limited number of systems have been studied, there seems to be no fundamental limitations to the application of Raman scattering from surface phonons. To favour the observation of surface phonons, the exciting laser light should be chosen such that surface resonant conditions are established while a small penetration of the light minimizes scattering contributions from bulk phonons. These conditions are of course most easily fulfilled in semiconducting materials but other systems, such as insulators and heteroterminated metal surfaces, for example, may also qualify.

Acknowledgements

We are very much indebted to M. Cardona, F. Bechstedt, G. Benedek, P.V. Santos, W.G. Schmidt, J. Fritsch and S. Lyapin for many helpful discussions concerning the topics of this article. Additional thanks are expressed to M. Cardona for careful reading and many useful suggestions during the course of manuscript writing. We are also grateful to P. Haier, K. Hinrichs and P. Marsiske for their assistance during the preparation of the manuscript.

Concerning our own contribution to the field of Raman scattering from surface phonons we would also like to express our sincere gratitude to the Deutsche Forschungsgemeinschaft (DFG) and the European Community (EU) which have provided financial support in several projects, namely Sfb202, Sfb6, Sfb290 (DFG) and EPIOPTICS, EASI (EU).

References

3.1 W. Hayes, R. Loudon: *Light Scattering in Solids* (Wiley, New York 1978)
3.2 M. Cardona: In *Light Scattering in Solids II*, ed. by M. Cardona, G. Güntherodt Topics Appl. Phys. **50**, (Springer, Berlin, Heidelberg 1982) p.19
3.3 B. Jusserand, M. Cardona: In *Light Scattering in Solids V*, ed. by M. Cardona, G. Güntherodt, Topics Appl. Phys. **66** (Springer, Berlin, Heidelberg 1989) p.49
3.4 J. Sapriel, B. Djafari Rouhani: Surf. Sci. Rep. **10**, 189 (1989)
3.5 F.H. Pollak: in *Analytical Raman Spectroscopy*, ed. by J.G. Grasseli, B.J. Bulkin, Chem. Anal. Ser. **114**, (Wiley, New York 1991)
3.6 A. Pinczuk, E. Burstein: In *Light Scattering in Solids*, ed. by M. Cardona, G. Güntherodt, Topics Appl. Phys. **8** (Springer, Berlin, Heidelberg 1975) p. 23

3.7 G. Abstreiter, M. Cardona, A. Pinczuk: In *Light Scattering in Solids IV*, ed. by M. Cardona, G. Güntherodt, Topics Appl. Phys. **54** (Springer, Berlin, Heidelberg 1984), p.5
3.8 K. Hinrichs, A. Schierhorn, P. Haier, N. Esser, W. Richter, J. Sahm: Phys. Rev. Lett. **79**, 1094 (1997)
3.9 A. Campion, J. Electron. Spectrosc. Relat. Phenom. **54/55**, 877 (1990)
3.10 A. Otto: In *Light Scattering in Solids IV*, ed. by M. Cardona, G. Güntherodt, Topics Appl. Phys. **54** (Springer, Berlin, Heidelberg 1984)
3.11 K. Arya, R. Zeyher: In *Light Scattering in Solids IV*, ed. by M. Cardona, G. Güntherodt Topics Appl. Phys. **54** (Springer, Berlin, Heidelberg 1984)
3.12 A. Otto: J. Raman Spectrosc. **22**, 743 (1991)
3.13 W. Suetaka: *Surface Infrared and Raman Spectroscopy, Methods and Applications* (Plenum, New York 1995)
3.14 W. Richter: In *Resonant Raman Scattering in Semiconductors*, ed. by G. Höhler: Springer Tracts Mod. Phys. **78** (Springer, Berlin, Heidelberg 1976)
3.15 G. Bauer, W. Richter (ed.): Optical Characterization of Epitaxial Semiconductor Layers (Springer, Berlin, Heidelberg 1996)
3.16 J.F. McGilp, D. Weaire, Ch. Patterson (eds.): *EPIOPTICS, Linear and Nonlinear Optical Spectroscopy of Surfaces and Interfaces* (Springer, Berlin, Heidelberg 1995)
3.17 F.W. de Wette: In *Surface Phonons*, W. Kress, F.W. De Wette (eds.) Springer Ser. Surf. Sci. **21** (Springer, Berlin, Heidelberg 1991)
3.18 W. Kress, F.W. De Wette (eds.): *Surface Phonons*, Springer Ser. Surf. Sci. **21** (Springer, Berline, Heidelberg 1991)
3.19 M. Lannoo, P. Friedel: *Atomic and Electronic Structure of Surfaces*, Springer Ser. Surf. Sci. **16** (Springer, 1991)
3.20 R.F. Wallis: Surf. Sci. bf 299/300 612 (1994)
3.21 G. Benedek, J.P. Toennies: Surf. Sci. **299/300**, 587 (1994)
3.22 Lord Rayleigh: Proc. London Math. Soc. **17**, 4 (1887)
3.23 M. Born, Th. von Karman: Z. Phys. **13**, 297 (1912)
3.24 I.M. Lifshitz, L.M. Rosenzweig: Zh. Eksp. Teor. Fiz. **18**, 1012 (1948) [Engl.Transl.: I.M. Lifshitz, Nuovo Cim. Suppl. **3**, 732 (1956)]
3.25 R. Fuchs, K.L. Kliever: Phys. Rev. A **140**, 2076 (1965)
3.26 R.F. Wallis: In *Atomic Structure and Properties of Solids*, ed. by E. Burstein (Academic, New York 1972)
3.27 A.A. Lucas: J. Chem. Phys. **48**, 3156 (1968)
3.28 R.E. Allen, G.P. Alldredge, F.W. de Wette: Phys. Rev. B **4**, 1661 (1971)
3.29 J. Fritsch, P. Pavone, U. Schröder: Phys. Rev. Lett. **71**, 4194 (1993)
3.30 W.G. Schmidt, F. Bechstedt, G.P. Srivastava: Phys. Rev. B **52** 2001 (1995) and references therein
3.31 B.N. Brockhouse, A.T. Stewart: Phys. Rev. **100**, 756 (1955)
3.32 K.H. Rieder, E.M. Hörl: Phys. Rev. Lett. **20**, 209 (1968)
3.33 K.H. Rieder, W. Drexel: Phys. Rev. Lett. **34**, 148 (1975)
3.34 H. Ibach, D.L. Mills: *Electron Energy Loss Spectroscopy and Surface Vibrations* (Academic, New York 1982)
3.35 H. Ibach: *Electron Energy Loss Spectrometers – The Technology of High Performance* (Springer, Berlin, Heidelberg 1991)
3.36 H. Ibach, M. Balden, D. Bruchmann, S. Lehwald: Surf. Sci. **269/270**, 94 (1992)

3.37 J. P. Toennies, K. Winkelmann: Chem. Phys. **66**, 3965 (1977)
3.38 J. P. Toennies: In *Surface Phonons*, Experimental Determination of Surface Phonons by Helium Atom and Electron Energy Loss Spectroscopy, W. Kress, F.W. De Wette (eds.) Springer Ser. Surf. Sci. **21** (Springer, Berlin, Heidelberg 1991)
3.39 Y.J. Chabal: Surf. Sci. Rep. **8**, 282 (1988)
3.40 D.E.Aspnes: In *Handbook of Semiconductors Vol. 2*, ed. by M. Balkanski (North Holland, Amsterdam 1980)
3.41 M. Born, K. Huang: *Dynamical Theory of Crystal Lattices* (Clarendon, Oxford 1966)
3.42 R. Zeyher, M. Cardona, H. Bilz: Sol. Stat. Commun. **19**, 57 (1976)
3.43 F. Stern: Phys. Rev. Lett. **18**, 546 (1967)
3.44 R. Loudon: Proc. Royal Soc. A **275**, 218 (1963)
3.45 A. Cantarero, C. Trallero-Giner, M. Cardona: Phys. Rev. B **39**, 8388 (1989)
3.46 E.O. Kane: Phys. Rev. **178**, 1368 (1969)
3.47 M. Born, K. Huang: *Dynamical Theory of Crystal Lattices* (Clarendon, Oxford, 1954)
3.48 P. Brüesch: In *Phonons: Theory and Experiments I*, ed. by M. Cardona, P. Fulde, H.J. Queisser, Springer Ser. Solid-State Sci. **34** (Springer, Berlin, Heidelberg 1982) pp. 117
3.49 H. Fröhlich: Adv. Phys. **3**, 325 (1954)
3.50 R. Trommer, M. Cardona: Phys. Rev. B **17**, 1865 (1978)
3.51 M. Sinyukov, R. Trommer, M. Cardona: Phys. Stat. Sol. (b) **86**, 563 (1978)
3.52 A. Compaan, H.J. Trodahl: Phys. Rev. B **29**, 793 (1984)
3.53 G.F. Koster, J.O. Dimmock, R.G. Wheeler, H. Statz: *Properties of the Thirty-two Point Groups* (M.I.T., Cambridge, MA 1963)
3.54 R. Loudon: Adv. Phys. **13**, 423 (1964)
3.55 R. Claus, L. Merten, J. Brandmüller: Springer Tracts Mod. Phys. **75**, G. Höhler (ed.) (Springer, Berlin, Heidelberg 1975)
3.56 R.K. Chang, M.B. Long: In [3.2] p. 179
3.57 J.C. Tsang: In *Light Scattering in Solids V*, ed. by M. Cardona, G. Güntherodt, Topics Appl. Phys. **66** (Springer, Berlin, Heidelberg 1989) p. 233
3.58 R.B. Bilhorn, J.V. Sweedler, P.M. Epperson, M.B. Denton: Appl. Spectrosc. **41**, 1114 (1987)
3.59 W.G. Schmidt, F. Bechstedt, G.P. Srivastava: Surf. Sci. Rep. **25**, 141 (1996)
3.60 N. Esser, J. Geurts: In *Optical Characterization of Epitaxial Semiconductor Layers*, ed. by G. Bauer, W. Richter (Springer, Berlin, Heidelberg 1996)
3.61 J.F. McGilp, A.B. McLean: J. Phys. C **21**, 807 (1988)
3.62 P. Skeath, C.Y. Su, I. Lindau, W.E. Spicer: J. Vac. Sci. Technol. **17**, 874 (1980)
3.63 J. Carelli, A. Kahn: Surf. Sci. **116** 380 (1982)
3.64 C.B. Duke, A. Paton, W.K. Ford, A. Kahn, J. Carelli: Phys. Rev. B **26** 803 (1982)
3.65 P. Skeath, C.Y. Su, W.A. Harrison, I. Lindau, W. Spicer: Phys. Rev. B **27**, 6246 (1983)
3.66 A. Kahn: Surf. Sci. Rep. **3** 193 (1983)
3.67 C.B. Duke, A. Patton: Surf. Sci. **164**, L 797 (1985)
3.68 W.K. Ford, T. Guo, D.L. Lessor, C.B. Duke: Phys. Rev. B **42**, 8952 (1990)

3.69 C.B. Duke, C. Mailhiot, A. Paton, K. Li, C. Bonapace, A. Kahn: Surf. Sci. **163**, 391 (1985)
3.70 W.K. Ford, T. Guo, K.-J. Wan, C.B. Duke: Phys. Rev. B **45**, 11896 (1992)
3.71 K. E. Miyano, T. Kendelewicz, J.C. Woicik, P.L. Cowan, C.E. Bouldin, B.A. Karlin, P. Pianetta, W.E. Spicer: Phys. Rev. B **46**, 6869 (1992)
3.72 T. Kendelewicz, J.C. Woicik, K.E. Miyano, A. Herrera-Gómez, P.L. Cowan, B.A. Karlin, C.E. Bouldin, P. Pianetta, W.E. Spicer: Phys. Rev. B **46**, 7276 (1992)
3.73 T. Kendelewicz, J.C. Woicik, K.E. Miyano, P.L. Cowan, B.A. Karlin, C.E. Bouldin, P. Pianetta, W.E. Spicer: J. Vac. Sci. Technol. B **9**, 2290 (1991)
3.74 M.G. Betti, C. Mariani, N. Jedrecy, R. Pinchaux, A. Ruocco, M. Sauvage-Simkin: Phys. Rev. B **50**, 14336 (1994)
3.75 C. Nowak, A. Chassé , A. Hempelmann, W. Richter, E. Dudzik, R. Whittle, I.T. McGovern, W. Braun, D.R.T. Zahn: Surf. Sci. **307–309**, 685 (1994)
3.76 C. Nowak: PhD thesis, TU Berlin (Wissenschaft & Technik, Berlin 1996)
3.77 C. Nowak, A. Hempelmann, A. Märkl, A. Chassé, E. Dudzik, C. Müller, I.T. McGovern, W. Braun, W. Richter, D.R.T. Zahn: Surf. Sci. **331–333**, 564 (1995)
3.78 C. Nowak, J. Krujatz, A. Märkl, C. Meyne, A. Chassé, W. Braun, W. Richter, D.R.T. Zahn: Surf. Sci. **331–333**, 619 (1995)
3.79 R.M. Feenstra, P. Mårtensson: Phys. Rev. Lett. **61**, (1988) 447
3.80 P. Mårtensson, R.M. Feenstra: Phys. Rev. B **39**, 7744 (1989)
3.81 P. Mårtensson, R.M. Feenstra: J. Microsc. **152**, 761 (1988)
3.82 R.M. Feenstra, P. Mårtensson, R. Ludeke: Mater. Res. Soc. Symp. Proc. **138**, 305 (1989)
3.83 C.K. Shih, R.M. Feenstra, P. Mårtensson: J. Vac. Sci. Technol. A **8**, 3379 (1990)
3.84 A.B. McLean, R.M. Feenstra, A. Taleb-Ibrahimi, R. Ludeke: Phys. Rev. B **39**, 12925 (1989)
3.85 F. Schäffler, R. Ludeke, A. Taleb-Ibrahimi, G. Hughes, D. Rieger: J. Vac. Sci. Technol. B **5**, 1048 (1987)
3.86 F. Schäffler, R. Ludeke, A. Taleb-Ibrahimi, G. Hughes, D. Rieger: Phys. Rev. B **36**, 1328 (1987)
3.87 R. Cao, K. Miyano, I. Lindau, W.E. Spicer: J. Vac. Sci. Technol. A **7**, 738 (1989)
3.88 R. Cao, K. Miyano, T. Kendelewicz, I. Lindau, W.E. Spicer: Surf. Sci. **206**, 413 (1988)
3.89 N. Esser, D.R.T. Zahn, C. Müller, W. Richter, C. Stephens, R. Whittle, I.T. McGovern, S. Kulkarni, W. Braun: Appl. Surf. Sci. **56–58**, 169 (1992)
3.90 R. Whittle, E. Dudzik, I.T. McGovern, A. Hempelmann, C. Nowak, D.R.T. Zahn, A. Cafolla, W. Braun: J. Electron Spectrosc. Relat. Phenom. **68**, 399 (1994)
3.91 R. Whittle, I.T. McGovern, D.R.T. Zahn, C. Müller, C. Nowak, A. Cafolla, W. Braun: Appl. Surf. Sci. **56–58**, 218 (1992)
3.92 A.B. McLean, D.M. Swanston, D.N. McIlroy, D. Heskett, R. Ludeke, H. Munekata: Phys. Rev. B **51**, 14271 (1995)
3.93 A. Tulke, H. Lüth: Surf. Sci. **178**, 131 (1986)
3.94 P. Mårtensson, G.V. Hansson, M. Lähdeniemi, K.O. Magnusson, S. Wiklund, J.M. Nicholls: Phys. Rev. B **33**, 7399 (1986)

3.95 A. Tulke, M. Mattern-Klosson, H. Lüth: Solid State Commun. **59**, 303 (1986)
3.96 C. Stringer, A. McKinley, G.J. Hughes, R.H. Williams: Vaccuum **33**, 597 (1983)
3.97 R.H. Williams, A. McKinley, G.J. Hughes, T.P. Humphreys, C. Maani: J. Vac. Sci. Technol. B **2**, 561 (1984)
3.98 W. Drube, F.J. Himpsel: Phys. Rev. B **37** 855 (1988)
3.99 F.J. Himpsel, W. Drube, A.B. McLean, A. Santoni: Appl. Surf. Sci. **56**, 160 (1992)
3.100 R. Strümpler, H. Lüth: Surf. Sci. **182**, 545 (1987)
3.101 M. Mattern-Klosson, R. Strümpler, H. Lüth: Phys. Rev. B **33**, 2559 (1986)
3.102 A.I. Shkrebtii, N. Esser, M. Köpp, P. Haier, W. Richter, R. Del Sole: Appl. Surf. Sci. **104/105**, 176 (1996)
3.103 N. Esser, M. Köpp, P. Haier, W. Richter: Phys. Stat. Solidi (b) **152**, 191 (1995)
3.104 C. Goletti, P. Chiaradia, W. Jian, G. Chiarotti: Solid State Commun. **84**, 421 (1992)
3.105 N. Esser, M. Köpp, P. Haier, A. Kelnberger, W. Richter: J. Vac. Sci. Technol. B **11**, 1481 (1993)
3.106 C. Mariani, G. Annovi, U. del Pennino, J. Electron Spectrosc. Relat. Phenom. **54/55**, 1105 (1990)
3.107 P. Chiaradia, A.I. Shkrebtii, C. Goletti, W. Jiang, R. Del Sole, Solid State Commun. **85**, 479 (1993); Solid State Commun. **89**, 87 (1994)
3.108 N. Esser, R. Hunger, J. Rumberg, W. Richter, R. Del Sole, A.I. Shkrebtii: Surf. Sci. **307–309**, 1045 (1994)
3.109 M.G. Betti, M. Pedio, U. del Pennino, C. Mariani: Phys. Rev. B **45**, 14057 (1992)
3.110 M.G. Betti, M. Pedio, U. del Pennino, C. Mariani: Phys. Rev. B **43**, 14317 (1991)
3.111 M.G. Betti, A. Campo, M. Capozi, M. Pedio, P. Perfetti, C. Quaresima, Vuoto XX, 75 (1990)
3.112 C. Mailhiot, C.B. Duke, D.J. Chadi: Phys. Rev. Lett. **53**, 2114 (1984)
3.113 C. Mailhiot, C.B. Duke, D.J. Chadi: J. Vac. Sci. Technol. A **3**, 915 (1985)
3.114 J.E. Northrup: Phys. Rev. B **44**, 1349 (1991)
3.115 J.P. LaFemina, C.B. Duke, C. Mailhiot: J. Vac. Sci. Technol. B **8**, 888 (1990)
3.116 G.P. Srivastava: Phys. Rev. B **46**, 7300 (1992)
3.117 G.P. Srivastava: Phys. Rev. B **47**, 16616 (1993)
3.118 G.P. Srivastava: Surf. Sci. **307–309**, 328 (1994)
3.119 W.G. Schmidt, B. Wenzien, F. Bechstedt: Phys. Rev. B **49**, 4731 (1994)
3.120 G.P. Srivastava: J. Phys.: Condens. Matter **5**, 4695 (1993)
3.121 C. Mailhiot, C.B. Duke, D.J. Chadi: Phys. Rev. B **31**, 2213 (1985)
3.122 C.M. Bertoni, C. Calandra, F. Manghi, E. Molinari: Phys. Rev. B **27**, 1251 (1983)
3.123 F. Manghi, C. Calandra, E. Molinari: Surf. Sci. **184**, 449 (1987)
3.124 G.P. Srivastava, R.P. Martin: J. Phys.: Condens. Matter **4**, 2009 (1992)
3.125 A. Bödicker, W. Schattke, J. Henk, R. Feder: J. Phys.: Condens. Matter **6**, 1927 (1994)
3.126 R. Whittle, I.T. McGovern, A. Hughes, T.-H. Shen, C.C. Matthai: J. Phys.: Condens. Matter **5**, 6555 (1993)
3.127 R.K. Ram, S.S. Kushwaha, A. Shukla: Phys. Stat. Solidi (b) **154**, 553 (1989)

3.128 G.F. Alfrey, P.H. Borcherds: J. Phys. C: Solid State Physics **5**, L275 (1972)
3.129 E. Bedel, G. Landa, R. Carles, J.P. Redoules, J.B. Renucci: J. Phys. C: Solid State Phys. **19**, 1471 (1986)
3.130 T. Sekine, K. Uchinokura, E. Matsuura: J. Phys. Chem. Solids **38**, 1091 (1977)
3.131 R. Trommer, M. Cardona: Phys. Rev. B **17**, 1865 (1978)
3.132 C. Patel, T.J. Parker, H. Jamshidi, W.F. Sherman: Phys. Stat. Sol. (b) **122**, 461 (1984)
3.133 D. Strauch, B. Dorner: J. Phys.: Condens. Matter **2**, 1457 (1990)
3.134 J. Fritsch, M. Arnold, U. Schröder: Phys. Rev. B (in press)
3.135 P. Giannozzi, S. de Gironcoli, P. Pavone, S. Baroni: Phys. Rev. B **43**, 7231 (1991)
3.136 P.V. Santos, N. Esser, J. Groenen, M. Cardona, W.G. Schmidt, F. Bechstedt: Phys. Rev. B **52**, 17379 (1995)
3.137 M. Hünermann, J. Geurts, W. Richter: Phys. Rev. Lett. **66**, 640 (1991)
3.138 W. Richter, N. Esser, A. Kelnberger, M. Köpp: Solid State Commun. **84**, 165 (1992)
3.139 N. Esser, M. Köpp, P. Haier, W. Richter: J. Electron Spectrosc. Relat. Phenom. **64/65**, 85 (1993)
3.140 N. Esser, W. Richter, U. Resch-Esser, P. Chiaradia, C. Goletti, L. Moretti: Philos. Mag. B **70**, 507 (1994)
3.141 P.V. Santos, N. Esser, M. Cardona, W.G. Schmidt, F. Bechstedt: Phys. Rev. B **52**, 12158 (1995)
3.142 N. Esser, M. Reckzügel, R. Srama, U. Resch, D.R.T. Zahn, W. Richter, C. Stephens, M. Hünermann: J. Vac. Sci. Technol. B **8**, 680 (1990)
3.143 J.C. Patrin, Y.Z. Li, M. Chander, J.H. Weaver: Phys. Rev. B **45**, 3918 (1992)
3.144 J.C. Patrin, Y.Z. Li, M. Chander, J.H. Weaver: Phys. Rev. B **46**, 10221 (1992)
3.145 P. Lautenschlager, M. Carriga, M. Cardona: Phys. Rev. B **36**, 4813 (1987)
3.146 M.K. Kelly, S. Zollner, M. Cardona: Surf. Sci. **285**, 282 (1993)
3.147 K. Hingerl, D.E. Aspnes, I. Kamiya, L.T. Florez: Appl. Phys. Lett. **63**, 885 (1993)
3.148 M. Kuball, M.K. Kelly, P.V. Santos, M. Cardona: Phys. Rev. B **50**, 8609 (1994)
3.149 W.G. Schmidt, G.P. Srivastava: Surf. Sci. **331–333**, 540 (1995)
3.150 K. E. Miyano, J.C. Woicik, T. Kendelewicz, W.E. Spicer, M. Richter, P. Pianetta: Phys. Rev. B **47**, 6444 (1993)
3.151 W.G. Schmidt, G.P. Srivastava: Solid State Commun. **89**, 345 (1994)
3.152 N. Esser, K. Hinrichs, J.R. Power, W. Richter: Surf. Sci. **427-428**, 44 (1999)
3.153 R.D.Bringans: Crit. Rev. Solid State Mat. Sci. **17**, 353 (1992)
3.154 R.S. Becker, B.S. Swartzentruber, J.S. Veckers, M.S. Hybertsen, S.G. Louie: Phys. Rev. Lett. **60**, 116 (1988)
3.155 J.R. Patel, J.A. Golovchenko, P.E. Freeland, H.J. Gossmann: Phys. Rev. B **36**, 7715 (1987)
3.156 M.S. Hybertsen, S.G. Louie: Phys. Rev. B **38**, 4033 (1988)
3.157 R.I.G. Uhrberg, R.D. Bringans, M.A. Olmstead, R.Z. Bachrach, J.E. Northrup: Phys. Rev. B **35**, 3945 (1987)
3.158 S. Bouzidi, T. Angot, F. Coletti, J.M. Debever, J.L.Guyaux, P.A.Thiry: Phys. Rev. B **49**, 16539 (1994)

3.159 R.B. Doak, D.B. Nguyen: Phys. Rev. B **41**, 3578 (1990)
3.160 P. Santini, P. Ruggerone, L. Miglio, R.B. Doak: Phys. Rev. B **46**, 9865 (1992)
3.161 D.C. Morse, E.J. Mele: Phys. Rev. B **40**, 3465 (1989)
3.162 J. Schmidt, H. Ibach: Phys. Rev. B **50**, 14354 (1994)
3.163 H. Wilhelm, W. Richter, U. Rossow, D.R.T. Zahn, D.A. Woolf, D.I. Westwood, R.H. Williams: Surf. Sci. **251/252**, 556 (1991)
3.164 D.E. Aspnes, A.A. Studna: Phys. Rev. B **27**, 985 (1983)
3.165 U. Rossow, H. Wilhelm, U. Frotscher, W. Richter, D.R.T. Zahn: Surf. Sci. **287/288**, 718 (1993)
3.166 R. Honke, P. Pavone, U. Schröder: Surf. Sci. **367**, 75 (1996)
3.167 P. Dumas, Y.J. Chabal, G.S. Higashi: Phys. Rev. Lett. **65**, 1124 (1990)
3.168 K. Hricovini, R. Günther, P. Thiry, A. Taleb-Ibrahimi, G. Indlekofer, J.E. Bonnet, P. Dumas, Y. Petroff, X. Blase, X. Zhu, S.G. Louie, Y.J. Chabal, P.A. Thiry: Phys. Rev. Lett. **70**, 1992 (1993)
3.169 M. Cardona: Phys. Stat. Solidi (b) **118**, 463 (1983)
3.170 S. Bouzidi, F. Coletti, J.M. Debever, P.A. Thiry, P. Dumas, Y.J. Chabal: Phys. Rev. B **45**, 1187 (1992)
3.171 P. Jakob, Y.J. Chabal: J. Chem. Phys. **95**, 2897 (1990)
3.172 P. Jakob, Y.J. Chabal: Appl. Phys. Lett. **59**, 2968 (1990)
3.173 P. Dumas, Y.J. Chabal: Chem. Phys. Lett. **181**, 537 (1991)
3.174 Ch. Stuhlmann, G. Bogdanyi, H. Ibach: Phys. Rev. B **45**, 6786 (1992)
3.175 M.A. Hines, T.D. Harris, A.L. Harris, Y.C. Chabal: J. Electron. Spectrosc. Rel. Phenom. **64/65**, 183 (1993);
a M.A. Hines, Y.J. Chabal, T.D. Harris, A.L. Harris: Phys. Rev. Lett. **71**, 2280 (1993);
b M.A. Hines, Y.J. Chabal, T.D. Harris, A.L. Harris: J. Chem. Phys. **101**, 8055 (1994)
3.176 K. Raghavachari, P. Jakob, Y.J. Chabal: Chem. Phys. Lett. **206**, 156 (1993)
3.177 H. Sano, S. Ushioda: Phys. Rev. B **53**, 1958 (1996)
3.178 W.M. Lau, S. Jin, X.W. Wu, S. Ingrey: J. Vac. Sci. Technol. A **9**, 994 (1991)
3.179 Y. Tao, A. Yelon, E. Sacher, Z.H. Lu, M.J. Graham: Appl. Phys. Lett. **60**, 2669 (1992)
3.180 C.J. Sandroff et al: Appl. Phys. Lett. **51**, 33 (1987)
3.181 J.M. Jin, M.W.C. Dharma-wardana, D.J. Lockwood, G.C. Aers, Z.H. Lu, L.J. Lewis: Phys. Rev. Lett. **75**, 878 (1995)
3.182 R. Merlin, C. Colvard, M.V. Klein, H. Morkoc, A.Y. Cho, A.C. Gossard: Appl. Phys. Lett. **36**, 43 (1980)
3.183 A.K. Sood, J. Menendez, M. Cardona, K. Ploog: Phys. Rev.Lett. **54**, 2111 (1985)
3.184 R.E. Camley, D.L. Mills: Phys. Rev. **29**, 1695 (1984)
3.185 A.K. Arora, A.K. Ramdas, M.R. Melloch, N. Otsuka: Phys. Rev. B **36**, 1021 (1987)
3.186 A.J. Shields et al.: Phys. Rev. Lett. **72**, 412 (1994)
3.187 M.P. Chamberlain, M. Cardona, B.K. Ridley: Phys. Rev. B **48**, 14356 (1993)
3.188 A. Fasolino, E. Molinari, J.C. Maan: Superlattices Microstruc. **3**, 117 (1987)
3.189 S.G. Lyapin, P.C. Klipstein, N.J. Mason, P.J. Walker: Phys. Rev. Lett. **74**, 3285 (1995)
3.190 I. Sela: Phys. Rev. B **46**, 7200 (1992)
3.191 I. Sela, C.R. Bolognesi, L.A. Samoska, H. Kroemer: Appl. Phys. Lett. **60**, 3283 (1992)

3.192 B.V. Shanabrook, B.R. Bennet, R.J. Wagner: Phys. Rev. B **48**, 17172 (1993)
3.193 J. Spitzer, H.D. Fuchs, P. Etchegoin, M. Ilg, M. Cardona, B. Brar, H. Kroemer: Appl. Phys. Lett. **62**, 2274 (1993)
3.194 J. Spitzer, A. Hoepner, M. Kuball, M. Cardona, B. Jenichen, H. Neuroth, B. Brar, H. Kroemer: J. Appl. Phys. **77**, 811 (1995)
3.195 S.G. Lyapin, P.C. Klipstein, N.J. Mason, P.J. Walker: Superlat. Microstruct. **15**, 499 (1994)
3.196 S.G. Lyapin, A.V. Lomsadze, I.A. Trojan, P.C. Klipstein, N.J. Mason, P.J. Walker: In *The Physics of Semiconductors*, Proc. 23rd Int. Conf. Physics of Semiconductors, 1996, ed. by M. Scheffler, R. Zimmermann (World Scientific, Singapore 1996)
3.197 K. Hinrichs, N. Esser, W. Richter: Il Nuovo Cimento D **20**, 1007 (1998)
3.198 K. Hinrichs: Phys. Stat. Solidi (a) **170**, 279 (1998)
3.199 C. Mailhiot, C.B. Duke, D.J. Chadi: Surf. Sci. **149**, 366 (1985)
3.200 J.L.A. Alves, J. Hebenstreit, M. Scheffler: Phys. Rev. B **44**, 6188 (1991)
3.201 H. Nienhaus, W. Mönch: Surf. Sci. Lett. **328**, (1995) L561
3.202 G. K. Schenter, J. P. LaFemina: J. Vac. Sci. Technol. A **10**(4), 2429 (1992)
3.203 J. Fritsch, P. Pavone, U. Schröder: Phys. Rev. B **52**, 11326 (1995)
3.204 H. M. Tütüncü, G.P. Srivastava: Phys. Rev. B **53**, 15675 (1996)
3.205 W. G. Schmidt, F. Bechstedt, G. P. Srivastava: Phys. Rev. B **52**, 2001 (1995)
3.206 A. Umerski, G. P. Srivastava: Phys. Rev. B **51**, 2334 (1995)
3.207 R. Fuchs, K. L. Kliewer: Phys. Rev. **140**, A2076 (1965)
3.208 S. Ushioda, R. Loudon: In *Modern Problems of Condensed Matter Sciences, Surface Polaritons, Chap. 12*, ed. by V.M. Agranovich, D.L. Mills (North Holland, Amsterdam 1982)
3.209 L. Koenders, F. Bartels, H. Ullrich, W. Mönch: J. Vac. Sci. Technol. B **3**(4), 1107 (1985)
3.210 H. Carstensen, R. Claessen, R. Manzke, M. Skibowski: Phys. Rev. B **41**, 9880 (1990)
3.211 F. Manghi, E. Molinari, C.M. Bertoni, C. Calandra: J. Phys. C **15**, 1099 (1982)
3.212 S. Selci, A. Cricenti, A.C. Felici, C. Goletti, G. Chiarotti: Phys. Rev. B **44**, 8372 (1991)
3.213 A. Cricenti, A.C. Felici, C. Goletti, G. Chiarotti: Surf. Sci. **211/212**, 552 (1989)
3.214 J. Fritsch, M. Arnold, C. Eckl, R. Honke, P. Pavone, U. Schröder: Surf. Sci. **427-428**, 58 (1999)

4 Coherent Phonons in Condensed Media

Thomas Dekorsy, Gyu Cheon Cho, and Heinrich Kurz

With 24 Figures

The impulsive excitation and phase-sensitive detection of coherent phonons and phonon–polaritons provide a detailed insight into the dynamical properties of matter. The experiments are based on optical pump–probe techniques with femtosecond time resolution. These techniques enable the detection of amplitude and phase of the coherent lattice motion simultaneously. Frequencies in the terahertz range and dephasing times in the picosecond range can be obtained with high accuracy. Especially in semiconductors and semiconductor heterostructures, where a coherent phonon mode and free carriers are excited simultaneously, important information about carrier–phonon interaction far away from equilibrium is obtained. This chapter presents an overview of recent achievements in this lively field of condensed-matter physics.

4.1 Introduction

The time-resolved detection of transient phenomena in the picosecond and subpicosecond range has been enabled by the development of ultrafast laser sources. Ultrafast laser systems allow us to perform pump–probe experiments in several configurations concerning the photon energy, the light intensity, polarization selectivity, etc., thus a huge variety of impulsively stimulated processes can be studied in condensed media. Ultrafast time-resolved optical techniques have contributed to disentangle the time constants of the most important interactions in semiconductors and semiconductor heterostructures. A thorough review of the whole field has been given recently by *Shah* [4.1].

First time-resolved observations of non-equilibrium phonon dynamics were performed by means of time-resolved Raman scattering techniques [4.2, 4.3, 4.4]. By time-resolving the intensity of the anti-Stokes line, these experiments yield information on the decay of coherently excited phonon populations. The requirements for time-resolving a Stokes or anti-Stokes Raman peak is that the investigations pulse width in the energy domain must be smaller than the phonon energy, i.e., the pulse duration must be larger than a phonon period. Otherwise the observed phonon line would lie within the spectrum of the laser, aggravating the detection of the Raman shifted intensity. Therefore these techniques do not allow the resolution of the coherent phonon amplitude phase sensitively. Extensive investigations of the phonon dynamics in InP and GaAs have been performed by this technique by *Vallée* et al. [4.5, 4.6, 4.7].

Topics in Applied Physics, Vol. 76
Light Scattering in Solids VIII
Eds.: M. Cardona, G. Güntherodt

For a review on coherent anti-Stokes Raman scattering techniques and results we refer to *Bron* [4.8].

Laser pulses with durations sufficiently shorter than the inverse of a fundamental lattice vibration frequency allow us to excite this lattice mode coherently, i.e., with a large number of phonons in one mode with constant phase relation [4.9]. This coherently excited mode is associated with a macroscopic lattice distortion. This chapter deals with the mechanisms responsible for the generation of coherent phonons and the different experimental possibilities for their detection. The focus lies on the excitation of opaque semiconductors. Analogies and differences to the well established techniques of Raman spectroscopy will be outlined. The experiments performed on coherent phonons within the last years have enlarged our knowledge of non-equilibrium states of phonons in a way complementary to the knowledge gained by the mature and highly sophisticated field of inelastic light scattering in solids, for which the contributions to this series are outstanding references.

One important difference between Raman spectroscopy and time-resolved coherent-phonon spectroscopy is the fact that in the latter the excitation and detection processes can be separated and that they may have different physical origins. In the detection process of coherent phonons, infrared-active and Raman-active media behave differently. In the case of polar lattices, a macroscopic polarization is generated by ir-active phonons at the phonon frequency, which leads to the emission of terahertz radiation [4.10, 4.11, 4.12]. Such a macroscopic polarization at phonon frequencies can be detected time resolved via nonlinear optical processes such as the first-order electro-optic effect (related to the second-order nonlinear susceptibility $\chi^{(2)}$) [4.13] and the Franz-Keldysh effect (related to $\chi^{(3)}$) [4.14] in bulk semiconductors. Detection processes based on the quantum confinement of electronic states become relevant in two-dimensional semiconductor heterostructures and quantum dots. For non-polar lattice modes, i.e., only Raman-active modes, the detection process is closely related to the Raman susceptibility. In this case, the macroscopic lattice distortion leads to a change in reflectivity or transmission based on the Raman tensor $(\partial\chi/\partial Q)Q$, where Q is the atomic displacement coordinate. While this detection is conceptually very close to inelastic light-scattering experiments, there exist some fundamental differences. Let us assume a time interval of at least one phonon period for the inelastic interaction of a photon and a phonon in order to impose a sideband at the phonon frequency $\pm\omega_{\mathrm{phonon}}$ on the light at frequency ω. A sufficiently short laser pulse can be regarded as a delta-function in time compared to the oscillation period of the phonon, thus it represents a stroboscopic illumination of the crystal in a certain coherent displacement state Q. The intensity changes of the transmitted or reflected probe pulse do not necessarily obey the restrictions of wavevector and energy conservation for an inelastic scattering process.

This chapter is organized as follows: The generation and detection processes for coherent phonons are outlined in Sect. 4.2 and Sect. 4.3, respectively. The coherent phonon generation, detection, and manipulation by femtosecond laser pulses is discussed in detail in Sect. 4.4 for the case of GaAs, since GaAs plays the role of a key material for the understanding of a non-Raman excitation process for coherent LO phonons in opaque semiconductors. Section 4.5 discusses the observation of coherent phonons and coupled phonon-carrier excitations in low-dimensional semiconductor heterostructures. Experiments performed on Te single crystals reveal the possibility to detect coherent phonons via their Raman activity or via their ir activity (Sect. 4.6). Thus it is shown that time-resolved optical pump–probe techniques and THz emission spectroscopy represent the counterpart to cw Raman and cw ir spectroscopy, respectively. High excitation experiments in Te allow the observation of an optically induced phase transition via monitoring the coherent-phonon frequency. Results on phonon–polaritons in ferroelectric crystals and high-temperature superconductors are briefly discussed in Sect. 4.7. Finally, recent achievements in the field of coherent-phonon spectroscopy are summarized in Sect. 4.8.

4.2 Coherent-Phonon Generation

In this section we present a phenomenological description of the excitation of coherent phonons. The equation of motion of coherent phonons can be expressed as a driven harmonic oscillator for the coherent-phonon displacement amplitude Q [4.2,4.15]

$$\mu^* \left[\frac{\partial^2 Q(t)}{\partial t^2} + 2\gamma_{\text{phonon}} \frac{\partial Q(t)}{\partial t} + \omega_{\text{phonon}}^2 Q(t) \right] = F^Q(t), \tag{4.1}$$

where μ^* is the reduced lattice mass, γ_{phonon} is a phenomenological damping constant, and $F^Q(t)$ the appropriate driving force. Equation (4.1) can be derived from a quantum-mechanical model for the coherent-phonon amplitude [4.9]. The damping constant γ_{phonon} is related to the dephasing time T_2 of the coherent mode via $\gamma_{\text{phonon}} = 1/T_2$. The dephasing time $2/T_2 = 1/T_1 + 1/T_\text{p}$ is a combination of phase-destroying contributions $\sim T_\text{p}$ and truly population-decreasing contributions $\sim T_1$ [4.16]. The latter are dominated by anharmonic decay processes, e.g., the decay of LO phonons into acoustic phonons. Other population-decreasing processes are based on electron–phonon interaction, e.g., the absorption of a phonon by an electron.

The driving force $F^Q(t)$ is exerted by an ultrashort laser pulse. Different types of driving forces have been classified into Raman-type and non-Raman-type excitation [4.17]. The Raman driving force is given in lowest order by the interaction of the electromagnetic field E and the Raman polarizability $(\partial\alpha)/(\partial Q)$ of the medium,

$$F^Q(t) = \frac{1}{2} \left(\frac{\partial \alpha}{\partial Q} \right)_{kl} E_k E_l \,. \tag{4.2}$$

Another useful distinction can be made into impulsive and resonant type of excitation. An impulsive excitation of the phonon mode is explained by a model denoted as Displacive Excitation of Coherent Phonons (DECP) [4.18, 4.19, 4.20], which is based on an interband excitation from bonding to antibonding orbitals. This process leads to an impulsive change of the equilibrium position of the atoms and thereby to a coherent excitation of lattice vibrations maintaining the crystal symmetry, i.e., A_1-modes for a large variety of materials. Displacively excited coherent phonons are characterized by a $\cos(\omega_{\text{phonon}} t)$ dependence of the coherent amplitude [4.20]. DECP has been treated in a more rigorous quantum-mechanical description by *Kuznetsov* and *Stanton* [4.9]. DECP is one of the most prominent examples for non-Raman excitation mechanisms of coherent phonons. Further non-Raman mechanisms will be discussed later. For a comprehensive distinction of Raman and non-Raman processes we also refer to the recent review by *Merlin* [4.17].

Impulsive resonant excitation can be accomplished by the combined action of two field components, E_k and E_l, delivered by the broad spectrum of the ultrashort laser pulse. According to wavevector- and energy-conservation rules a lattice mode with frequency $\omega_l - \omega_k = \omega_{\text{phonon}}$ and wavevector $\boldsymbol{k}_l - \boldsymbol{k}_k = \boldsymbol{k}_{\text{phonon}}$ is excited. This process is denoted as Impulsive Stimulated Raman Scattering (ISRS) [4.21, 4.22, 4.23, 4.24, 4.25]. The phonon wavevector can be manipulated by using two excitation pulses with well-defined angles of incidence. The initial phase of the coherent lattice displacement excited via ISRS should obey a $\sin(\omega_{\text{phonon}} t)$ dependence, however, *resonant* ISRS may also exhibit a $\cos(\omega_{\text{phonon}} t)$ dependence [4.25]. ISRS is one prominent Raman mechanism for the excitation of coherent phonons, which has been exploited most frequently for the generation of coherent phonons in the past.

For polar Raman-active LO phonons the driving force in (4.1) can be described in the following way [4.15]:

$$F_j^Q(t) = R_{jkl} E_k E_l - \frac{e^*}{\epsilon_\infty \epsilon_0} P_j^{\text{NL}} , \tag{4.3}$$

where R_{ijk} is the Raman tensor and P_j^{NL} is a nonlinear longitudinal polarization along direction j. The nonlinear polarization P_j^{NL} can be divided into several possible contributions [4.26]

$$\begin{aligned} P_j^{\text{NL}} \; &= \; \chi_{jkl}^{(2)} E_k E_l + \chi_{jklm}^{(3)} E_k E_l E_m \\ &\quad + \int_{-\infty}^{t} \mathrm{d}t' J_j(t') + Ne \int_{-\infty}^{t} \mathrm{d}t' \int_{-\infty}^{\infty} \mathrm{d}x_j \langle \Psi(x_j, t') | x_j | \Psi(x_j, t') \rangle \; , \end{aligned} \tag{4.4}$$

where the first two terms represent higher-order nonlinear susceptibilities, e.g., field-induced Raman scattering. The third and fourth terms describe polarizations set up by longitudinal intraband currents. The third term represents a longitudinal polarization generated by non-equilibrium currents with rise-times faster than the phonon period. They can impulsively excite coherent LO phonons in a current surge model for the screening of surface fields of III–V semiconductors [4.13, 4.27, 4.28, 4.11] (see Sect. 4.4.1) or via the ultrafast build-up of electric Dember fields [4.29, 4.10, 4.30] (see Sect. 4.6.2). It should be noted that these ultrafast currents do not require coherence in the electronic system. The last term of (4.4) describes the case of coherent electronic wavefunctions, which may set up a macroscopic intraband polarization. If the level splitting of coherently superposed electronic excitations is close to the LO-phonon resonance, i.e., the frequency of the intraband polarization is close to the phonon frequency, a direct interaction of the electronic polarization with the lattice polarization is expected [4.31].

For polar phonons the coupling of the lattice polarization given by a displacement Q and an electronic polarization P is given by a set of coupled equations [4.11]:

$$\frac{\partial^2}{\partial t^2}P_j + \gamma_{\mathrm{el}}\frac{\partial}{\partial t}P_j + \omega_{\mathrm{el}}^2 P_j = \frac{e^2 N}{\epsilon_\infty \mu^*}\left(E_j^{\mathrm{ext}} - 4\pi\gamma_{12}Q_j\right) \tag{4.5}$$

$$\frac{\partial^2}{\partial t^2}Q_j + \gamma_{\mathrm{phonon}}\frac{\partial}{\partial t}Q_j + \omega_{\mathrm{phonon}}^2 Q = \frac{\gamma_{12}}{\epsilon_\infty}\left(E_j^{\mathrm{ext}} - 4\pi P_j\right). \tag{4.6}$$

E_j^{ext} is a macroscopic electric field which has to be determined self-consistently via the Poisson equation and describes, e.g., the surface-field dynamics; γ_{12} is a coupling constant given by $\gamma_{12} = \omega_{\mathrm{TO}}\sqrt{(\epsilon_0 - \epsilon_\infty)/4\pi}$. The generalized phenomenological damping constant γ_{el} describes the damping of the electronic polarization, e.g., the momentum scattering time of a dense carrier plasma or the dephasing time of the intraband polarization of coherent electronic wavepackets. The frequency ω_{el} is either the plasma frequency or the frequency of a coherent wavepacket oscillation. The driving force in (4.5) and (4.6) is set up by a combined action of the electric-field dynamics and the polarization, which has been expressed in the different possible contributions in (4.3, 4.4).

Equations (4.5) and (4.6) describe coupled electron–phonon dynamics in a macroscopic sense. For the case of homogeneous densities and negligible damping, (4.5–6) reproduce the well-known dispersion branches of coupled plasmon–phonon modes for a one-component plasma [4.32]

$$\omega_{\mathrm{PPM}}^{\pm} = \frac{1}{2}\left(\omega_{\mathrm{LO}}^2 + \omega_{\mathrm{P}}^2 \pm \sqrt{(\omega_{\mathrm{P}}^2 + \omega_{\mathrm{LO}}^2)^2 - 4\omega_{\mathrm{P}}^2\omega_{\mathrm{TO}}^2}\right)^{(1/2)}, \tag{4.7}$$

with the plasma frequency $\omega_{\mathrm{P}}^2 = e_0^2 N/\epsilon_0\epsilon_\infty m^*$, N being the electron density and m^* the effective mass. The associated plasmon–phonon coupled modes were first observed in Raman experiments in doped GaAs [4.33, 4.34]. In order to temporally resolve coherent plasmon–phonon dynamics we have to bear in mind that an optically excited plasma may be strongly inhomogeneous in real space and k-space. For minimizing the lateral real-space inhomogeneity due to the excitation with a Gaussian beam profile it is incumbent to use excitation spots larger than the probe spot size. The spatial inhomogeneity perpendicular to the excited surface in absorbing materials has also to be considered.

The wavevector of the excited phonon modes depends strongly on the excitation process. For experiments in transparent media when the coherent-phonon excitation is based on first-order Raman processes, the wavevector is given by the wavelength λ of the pump laser, i.e., $k = 2\pi n/\lambda$, where n is the refractive index. In two-beam excitation configurations based on $\chi^{(2)}$ processes the coherent-phonon wavevector can be selected via the angle between the two laser pulses (see Sect. 4.7.2). For opaque semiconductors where the excitation is mainly based on non-Raman processes, e.g., DECP or the field screening mechanism, the excited phonon wavevector is mainly determined by the imaginary part of the refractive index κ, i.e., $k = 2\pi\kappa/\lambda$. Since the dispersion of optical phonons is weak for small wavevectors, their k cannot be determined from the observed frequency with high accuracy. This is not the case for acoustic zone-folded phonons in semiconductor superlattices (see Sect. 4.5.3) and acoustic modes in bulk materials excited at the edge of the Brillouin zone [4.35], where the frequencies observed allow one to disentangle details of the excitation mechanism.

4.3 Detection of Coherent Phonons

There are various possibilities to detect coherent phonons in the time domain and two main classes of detection methods can be distinguished. One class is based on the detection of changes in the optical properties of the investigated material at the frequency of the laser light. The other class is based on the time-resolved detection of far-infrared light in the terahertz frequency range

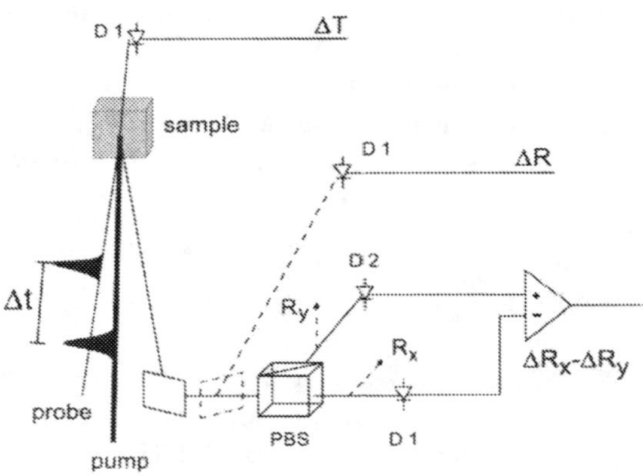

Fig. 4.1. Setup for the time-resolved all-optical detection of coherent phonons in transient reflectivity $\Delta R(\Delta t)$, transmission $\Delta T(\Delta t)$, or polarization-analyzed reflectivity changes $\Delta R_x(\Delta t) - \Delta R_y(\Delta t)$. D_1 and D_2 denote photo detectors and PBS a polarizing beam splitter

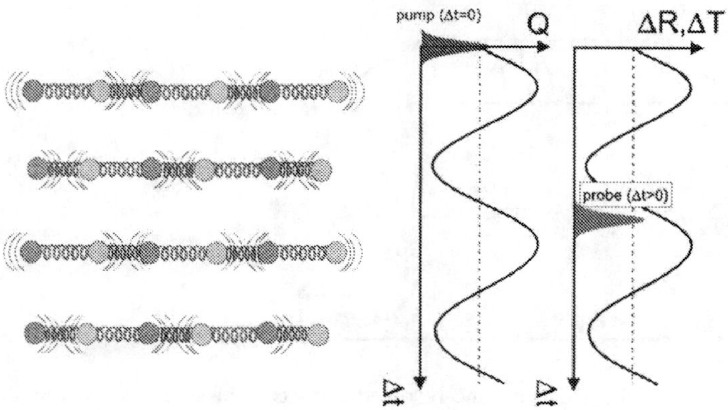

Fig. 4.2. Schematic sketch of the principle of time-resolved coherent-phonon detection. The pump pulse induces a coherent atomic displacement $Q(\Delta t)$, which is stroboscopically sampled in the intensity changes of the reflected or transmitted probe pulse

generated by coherent ir-active phonons. These two classes reflect the two historic continuous-wave methods for investigating phonon resonances, i.e., Raman spectroscopy and ir spectroscopy, respectively.

The first class can be divided into the time-resolved detection of reflected, transmitted or diffracted light. The mechanisms giving rise to a modulation of the optical properties at the phonon frequencies can themselves be manifold. Figures 4.1 and 4.2 depict schematically the experimental setup and the principle for detecting coherent phonons in the time domain, respectively.

The modulation of the reflectivity ΔR can be explained on the basis of the first-order Raman tensor $(\partial\chi)/(\partial Q)$, i.e.,

$$\Delta R = \frac{\partial R}{\partial n}\Delta n \approx \frac{\partial R}{\partial \chi}\frac{\partial \chi}{\partial Q}Q. \tag{4.8}$$

Since $(\partial\chi)/(\partial Q)$ is a tensor, different modes can be probed selectively by means of polarization analysis of the probe pulse [4.10, 4.36, 4.37]. For fully symmetric phonons the Raman tensor contains diagonal contributions of equal magnitude only (at least in a cubic crystal). In this case the reflectivity changes will be isotropic irrespective of the polarization of the probe pulse relative to the crystal orientation. For phonon modes with non-diagonal terms in the Raman tensor, $(\partial\chi)/(\partial Q)_{jk} \neq 0$ for $j \neq k$, the reflectivity changes will depend on the crystal orientation and the light polarization. Further detection schemes are based on Raman interaction with transmitted probe pulses, which results in a periodic shift of the spectral components of the probe pulse with the frequency of the lattice vibrations [4.38, 4.17]. In crystals exhibiting a linear electro-optic effect (Pockels effect), coherent phonons may be detected sensitively via the associated longitudinal field [4.13]. In terms of Raman tensors, the so-called electro-optic contribution to the Ra-

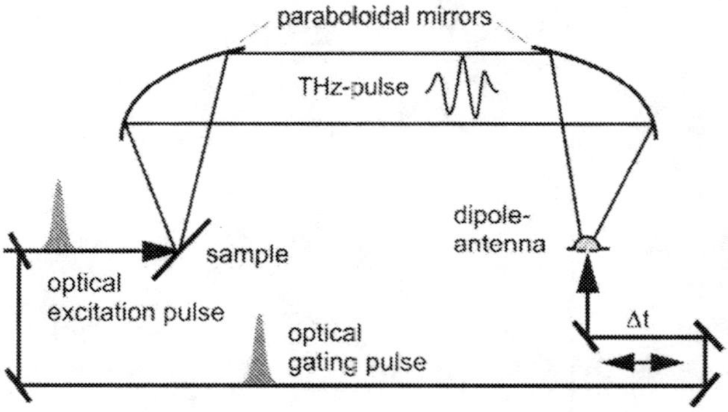

Fig. 4.3. Experimental setup for the time-resolved detection of terahertz electromagnetic waves

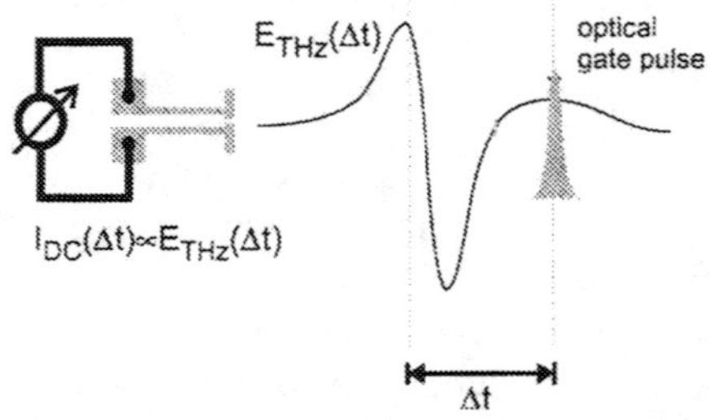

Fig. 4.4. Principle of the detection of terahertz (THz) electromagnetic pulses by using a gateable dipole antenna

man tensor $(\partial\chi/\partial F)(\partial F/\partial Q)Q$ has to be added to (4.8). The electric-field-induced phonon detection may be resonantly enhanced via third-order nonlinearities associated with the Franz-Keldysh effect close to interband resonances [4.39, 4.14].

Another important possibility for detecting coherent ir-active phonons, phonon–polaritons or plasmon–phonon modes is via the emissionemission of electromagnetic radiation in the terahertz frequency range. The emission characteristics can be assumed to follow the emission of a dipole, which is set up by the volume in which coherent phonons form a macroscopic polarization. The Poynting vector of the radiation is given by [4.11]

$$\boldsymbol{S}(t) = \frac{\omega_{\text{phonon}}^4 \boldsymbol{P}^2(t) \sin^2\theta}{16\pi^2\epsilon_0 r^2 c^3}\boldsymbol{s}, \tag{4.9}$$

where θ is the angle between the direction of observation $\boldsymbol{s}$ and the polarization $\boldsymbol{P}$, and r is the distance to the detector. Experimentally, the time-resolved detection of THz electromagnetic fields has become possible by the application of ultrafast photoconductive switches [4.40]. Figures 4.3 and 4.4 depict the experimental arrangement for time-resolving terahertz electromagnetic fields. The emitted THz radiation is collected by paraboloidal mirrors onto the dipole antenna which is activated by the time-delayed gating pulse.

The antenna, of approximately 50 μm dipole length, is fabricated on a semiconductor exhibiting free-carrier lifetimes in the subpicosecond range, e.g., ion-implanted silicon on sapphire or low-temperature-grown GaAs. The incident THz electric field drives a current in the outer circuit of the antenna during the time span of the gate pulse. The measured current is directly proportional to the incident field strength. Thus the THz electric field can be measured in amplitude and phase. The bandwidth of the antennas is roughly given by the free-carrier lifetime in the photoconductive switch (< 1 ps depending on the material used). Useful frequency bandwidths are typically in the range of 100 GHz to 3 THz. This limit excludes many interesting phonon modes in semiconductors to be studied. However, recent developments in the detection of terahertz radiation via electro-optic crystals are highly promising for achieving a detection bandwidth up to 30 THz [4.41, 4.42].

4.4 Coherent LO Phonons in GaAs

4.4.1 Coherent-Phonon Generation and Detection in GaAs

The coherent-phonon generation in opaque semiconductors was first observed by *Cho* et al. in the time-resolved reflectivity changes in GaAs [4.13] and by *Cheng* et al. in Bi and Sb [4.18]. Here we focus on the experiments in GaAs, a material which has been intensively studied by Raman spectroscopy [4.43, 4.44]. The results obtained by fs spectroscopy on GaAs clearly demonstrate some of the main points of time-resolved phonon spectroscopy, i.e.,

- the impulsive excitation via non-Raman effects due to the excitation of a dense electron–hole plasma in combination with built-in electric fields, i.e., the dynamics described by (4.5, 6),
- the visibility of coherent phonons via $\partial\chi/\partial Q$ according to the electro-optic Raman tensor and higher-order terms,
- the dispersion of the detection process in the vicinity of interband critical points, e.g., the resonance enhancement of the Franz-Keldysh effect,
- the coupling between coherent LO phonons and non-equilibrium carrier distributions.

The laser system employed in the first subpicosecond time-resolved experiments on GaAs was a Colliding-Pulse Mode-locked (CPM) dye laser which delivers light pulses at 2 eV with pulse durations of 50 fs. For the detection of coherent phonons, which often give small contributions on a large electronic background signal, a special fast-scanning detection system has been developed [4.45, 4.13]. This detection system is based on scanning the time delay between pump and probe pulses with a retroreflector mounted onto an activated vibrator. This allows the scanning of the time delay at frequencies around 100 Hz. With a real-time data-acquisition system based on a home-built VME-bus computer system the data are accumulated without using

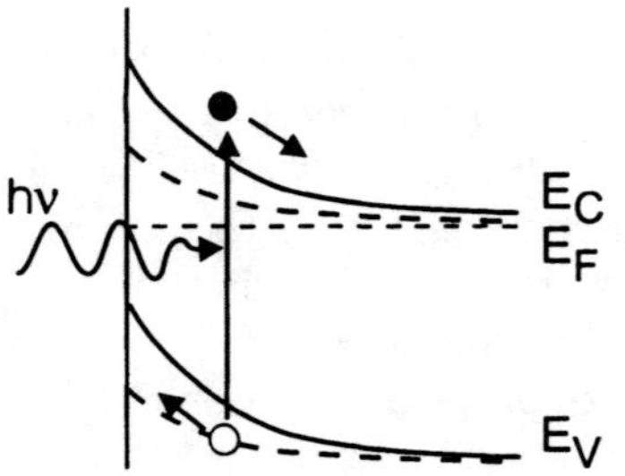

Fig. 4.5. Sketch of the field screening process at a bare surface of an n-doped semiconductor with mid-gap Fermi-level pinning at the surface. The solid (dashed) lines represent the band bending before (after) the optical excitation

further lock-in filtering techniques. This technique enabled the detection of time-resolved signal changes down to some 10^{-7} even with the CPM dye laser as pulse source [4.46].

In order to explain the excitation mechanism for coherent phonons in GaAs, electric surface fields at bare surfaces of III–V compounds have to be taken into account. Charged surface states lead to a pinning of the Fermi level within the gap, in turn producing a bending of the band structure towards the surface. The associated electric fields are calculated within a Schottky-barrier model [4.47] and exhibit a square-root dependence on the doping density of the crystal. An external manipulation of the surface field is possible via transparent Schottky contacts. Figure 4.5 sketches the underlying process for the generation of coherent LO phonons in GaAs surface space-charge fields. The optical injection of carriers within the surface-field region leads to an ultrafast current surge which rapidly screens the built-in electric field [4.28].

Associated with this ultrafast polarization change is the emission of broad-band terahertz radiation [4.40]. The ultrafast depolarization of the surface depletion field leads to the coherent excitation of LO phonons [4.13,4.27]. The detection for wavelengths far from resonances is accomplished via the Pockels effect [4.48]. On (100) GaAs surfaces, a longitudinal field along the [100]-direction leads to induced birefringence in the (100)-plane. By subtracting two polarization components of a linearly polarized probe pulse along the [011]- and [01$\bar{1}$]-direction, the following reflectivity change is [4.49, 4.13]

$$\frac{\Delta R(t)}{R_0} = \frac{\Delta R_{[011]}(t) - \Delta R_{[01\bar{1}]}(t)}{R_0} = \frac{4\,r_{41}\,n_0^3}{n_0^2 - 1}\,\Delta E_{\mathrm{surface}}(t)\,, \tag{4.10}$$

where r_{41} is the electro-optic coefficient, n_0 the unperturbed refractive index, and $\Delta E_{\mathrm{surface}}(t)$ the time-dependent macroscopic electric surface field along the [100]-direction.

Figure 4.6 depicts the electro-optic reflectivity changes of an n-doped GaAs sample prepared with a transparent indium–tin oxide Schottky contact on top. The optical excitation density is $4 \times 10^{17}\mathrm{cm}^{-3}$, the experiments were performed at room temperature. The transparent Schottky contact allows the change of the electric surface field by applying different voltages to the sample. The electro-optic measurements in Fig. 4.6 reveal the transient screening of the surface field, $\Delta E_{\mathrm{surface}}(t)$,for different initial surface

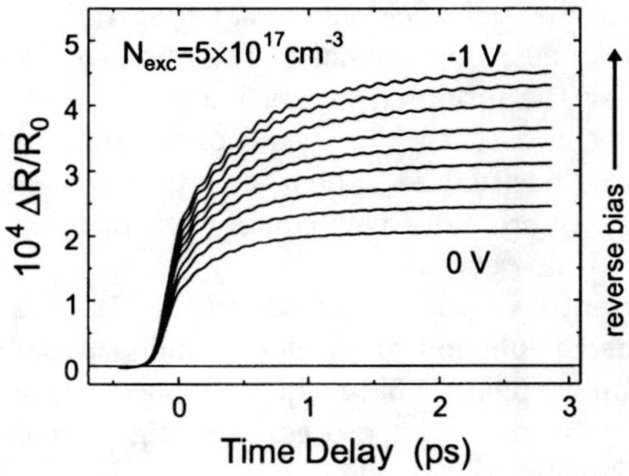

Fig. 4.6. Electro-optic reflectivity changes from (100) GaAs with a transparent Schottky contact on top. The data are obtained for different reverse-bias voltages at 300 K (from [4.28])

field strengths. The screening dynamics exhibits a fast component on the time scale of the exciting pulse, i.e., within one optical phonon period, and a slower component on a ps time scale. The associated electric field changes are on the order of 100 kV/cm [4.28]. The data are clearly modulated with oscillations with a frequency of 8.75 THz, matching the GaAs $\boldsymbol{k} = 0$ LO-phonon frequency at 300 K. The dephasing time of the oscillations, 4.0 ± 0.3 ps, is independent of the applied electric fields. This dephasing time is in agreement with the phonon lifetime derived from CARS experiments [4.5]. This agreement suggests that the observed decay of the coherent amplitude is determined by anharmonic decay and not by pure dephasing processes. Such an agreement between the dephasing time derived from coherent-phonon experiments and the linewidth obtained in Raman scattering experiments has also recently been observed in a thorough study of the dephasing in single-crystal bismuth [4.50].

In order to verify that the field screening is the responsible driving force of the coherent LO phonons, we plot the phonon amplitude versus the reflectivity change at a time delay of 50 fs (Fig. 4.7). This value is proportional to the driving force for coherent LO phonons according to the third term in (4.4),

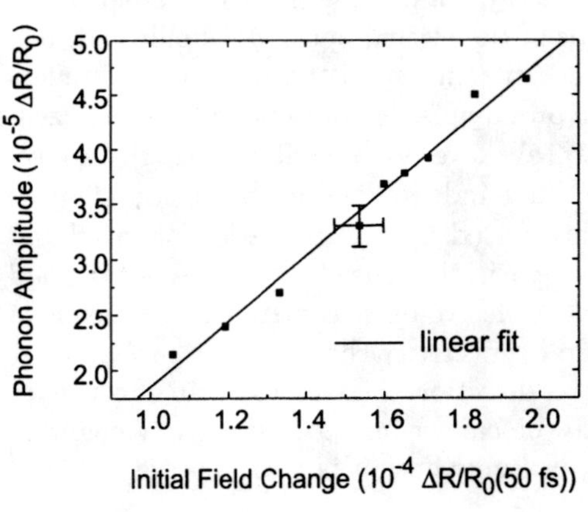

Fig. 4.7. Amplitude of coherent LO phonons (solid squares) versus the electro-optic reflectivity changes in Fig. 4.6 at a time delay of 50 fs (from [4.27])

since it represents the optically detected surface-field changes, i.e., $\Delta R(50\,\mathrm{fs}) \propto \Delta E_{\mathrm{surface}}(50\,\mathrm{fs})$. The data reveal a linear dependence of the phonon amplitude on $\Delta R(50\,\mathrm{fs})$, thus confirming the proposed generation mechanism. The initial phase of the oscillations is $\propto \cos(\omega t)$. This behavior is expected for an oscillator starting at a displaced position, i.e., the atomic positions at negative time delays, driven to oscillate around a new equilibrium position given by the electric field at positive time delays.

The excitation process of coherent LO phonons observed in GaAs is *isotropic* in the sense that the coherent-phonon amplitude is independent of the incident polarization of the pump pulse [4.38, 4.27]. This observation rules out a coherent phonon generation via a $\chi^{(2)}$ process [first right-hand term in (4.4)]. The last remaining Raman process which could, in principle, account for an isotropic generation process is a third-order nonlinear polarization with the third field being the surface field [second right-hand term in (4.4)]. This effect has been observed in cw Raman experiments close to electronic resonances and has been denoted as inverse Franz-Keldysh effect [4.51]. This effect has been shown to drop in scattering efficiency by orders of magnitude when the laser energy is detuned from interband resonances by a few tens of an meV. This effect could not be ruled-out completely as generation mechanism at the time when only femtosecond lasers with 2 eV photon energy were available. Therefore this question will be addressed later in the context of resonant excitation with a tunable femtosecond Ti : sapphire laser.

Intriguing situations in light-scattering experiments arise when the photon energy is in the vicinity of interband resonances [4.52, 4.53, 4.39]. The invention of the Kerr-lens mode-locked Ti:sapphire laser allows one now to perform coherent-phonon experiments over a wide range of excitation wavelengths (from 700 nm to 1000 nm). This range covers the bandgap of GaAs (1.42 eV at room temperature). Hence, different contributions to the resonant and non-resonant excitation and detection of coherent phonons can be investigated.

Figure 4.8 shows the time-resolved reflectivity change from a (100) GaAs surface for 2.0 eV and 1.47 eV laser energy obtained with two different probe polarizations along the principle axis of the electro-optic index ellipsoid, i.e., the [011] and [01$\bar{1}$] crystal directions. In contrast to the electro-optic reflectivity changes shown in Fig. 4.6, the probe pulse is not polarization analyzed. The observed reflectivity changes at 2 eV have two contributions: (*i*) An *electronic* contribution due to the refractive-index changes/absorption changes associated with the free carriers introduced by the excitation of interband transitions. This contribution is *isotropic* in the sense that it does not depend on the probe-pulse orientation. (*ii*) The electro-optic contribution, which is of comparable magnitude as the isotropic reflectivity change. For a probe polarization along the [011] ([01$\bar{1}$]) crystal direction the electro-optic effect gives a positive (negative) contribution to the carrier-induced reflectivity changes. This sign reversal of the electro-optic contribution is also observed for the

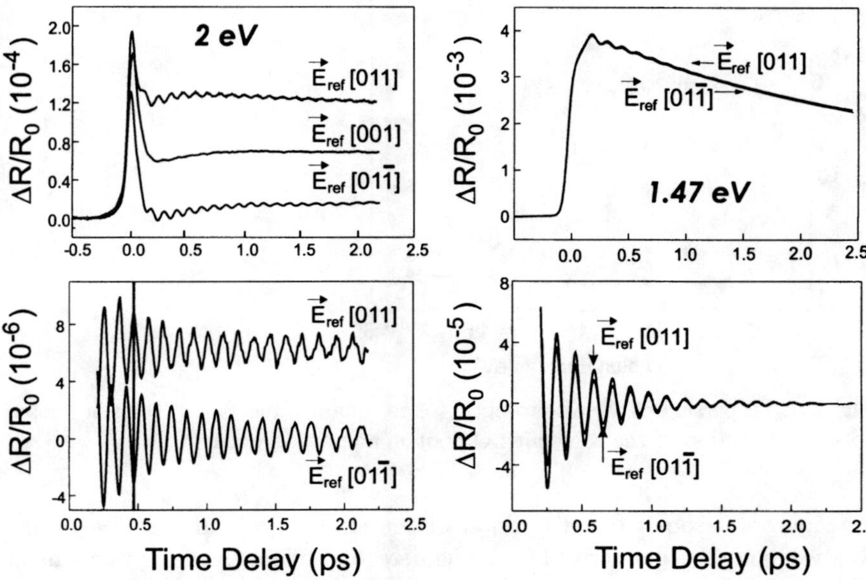

Fig. 4.8. Comparison of reflectivity changes of (100) GaAs at 2 eV (left) and 1.47 eV (right) photon energy for different polarizations of the reflected probe beam $\boldsymbol{E}_{\mathrm{ref}}$, i.e., along the [001], the [011], and the [01$\bar{1}$] direction. In the lower part the oscillatory contributions are compared for the [011] and the [01$\bar{1}$] polarization

phonon-induced contribution, as can be seen in the π phase shift between the oscillations. The same measurements performed close to the band edge exhibit a much larger total reflectivity change and a phonon induced contribution larger by more than an order of magnitude than the electro-optic contribution at 2 eV. In addition, the phase of the oscillations does not change with the probe polarization. Only a tiny difference exists between the two curves with a magnitude comparable to the difference between the two equivalent 2 eV curves. All these observations point toward the fact that the phonon detection at the fundamental bandgap is not based on the electro-optic effect but on an isotropic optical nonlinearity. This contribution arises from a third-order nonlinearity associated with the Franz-Keldysh effect [4.14]. The term in the expansion of the nonlinear third-order polarization responsible for this detection process is

$$P(\omega) = \epsilon \left[\chi^{(3)} E_{\mathrm{surface}}(0) E_{\mathrm{phonon}}(\omega_{\mathrm{LO}}) E(\omega) \right] , \tag{4.11}$$

where $E_{\mathrm{surface}}(0)$ is the surface field, $E_{\mathrm{phonon}}(\omega_{\mathrm{LO}})$ the field associated with the coherent LO phonons, and $E(\omega)$ the light field. The surface field $E_{\mathrm{surface}}(0)$ is assumed to be nearly static with respect to the light frequency. However, the subpicosecond surface-field dynamics also influences the coherent-phonon amplitude detected via (4.11).

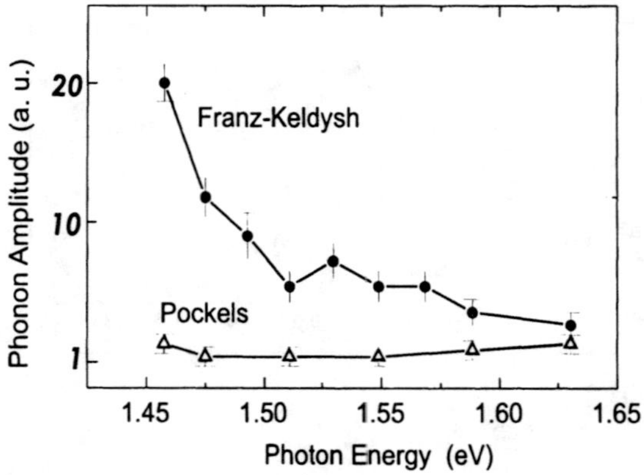

Fig. 4.9. Dispersion of the electro-optic (open squares) and Franz-Keldysh (closed dots) contribution to the coherent LO phonon amplitude (from [4.14]).

Figure 4.9 shows the dispersion of the electro-optic and of the Franz-Keldysh contribution derived from the detection of coherent LO-phonons at different laser energies, i.e., the dispersion of $\chi^{(2)}$ and $\chi^{(3)}$ at the LO-phonon frequency. The data reveal a nearly dispersion-free electro-optic effect, while the Franz-Keldysh-like detection is strongly enhanced when the energy approaches the band edge. This data give clear evidence for higher-order nonlinearities being relevant in the resonant detection of phonons. A dispersion of the generation process can be ruled out experimentally, since it would lead to a change in the phonon amplitude in the electro-optic detection scheme. Thus these measurements rule out the inverse Frank-Keldysh effect discussed above as a relevant excitation mechanism.

4.4.2 Coupled Plasmon–Phonon Modes

The interaction of phonons and plasmons has been an intensively investigated subject in Raman experiments since the first observations in doped semiconductors by *Mooradian* and *McWorther* [4.34]. The origin of such coupled modes has already been discussed in Sect. 4.2. Time-resolved experiments allow the study of non-equilibrium effects on the coupled dynamics. We focus on experiments in GaAs, where coherent LO phonons are generated and detected as discussed in the previous section.

Figure 4.10 shows early data on coupled plasmon–phonon modes obtained in the electro-optic detection configuration with 2 eV laser pulses delivered by a CPM laser [4.54]. The time-domain data exhibit a pronounced beating which is identified as a superposition of a coherent LO phonon and a broad plasmon contribution centered around the TO phonon, i.e., the screened LO mode. No distinct modes other than LO or screened LO could be observed. This fact is a shortcoming of the output power delivered from the CPM laser. In order to achieve high carrier densities in excess of $5\times10^{17}\text{cm}^{-3}$, necessary

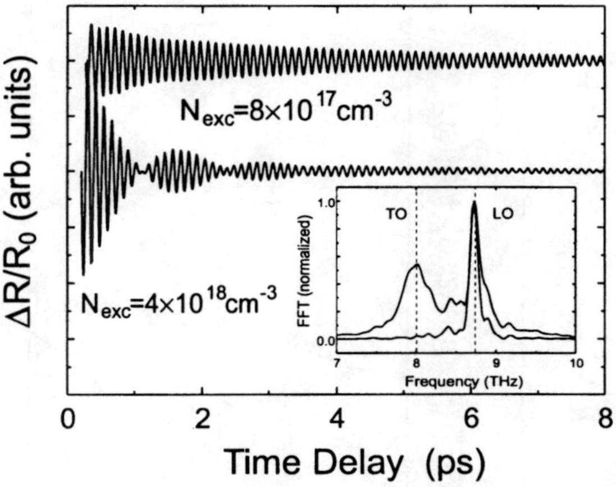

Fig. 4.10. Extracted phonon signature of electro-optic reflectivity changes obtained on (100) GaAs at 2 eV excitation for the two different densities given in the figure. The inset shows the Fourier transforms of the time-domain data

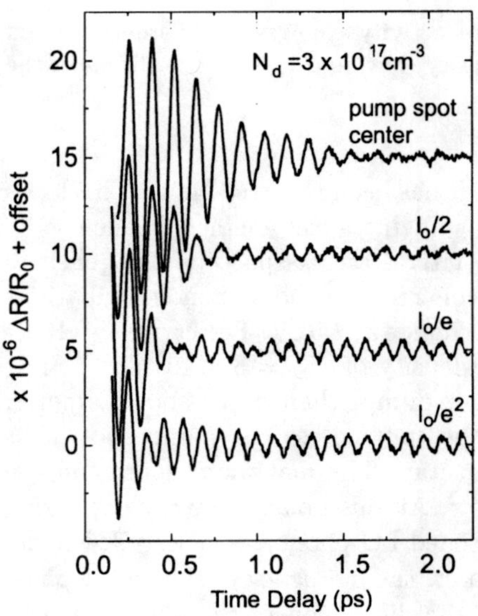

Fig. 4.11. Time-resolved phonon traces probed via the Pockels effect with a small probe spot scanned over a large Gaussian excitation spot (from [4.56])

to drive the plasma frequency into resonance with the LO phonon, the pump spot has to be focused tightly to $< 20\,\mu\text{m}$. At this focus size, the probe pulse averages over a lateral highly inhomogeneous carrier density, so that distinct plasmon modes cannot be resolved. Beside this spatial inhomogeneity, the optically generated plasma is also highly inhomogeneous in $\boldsymbol{k}$-space. At 2 eV excitation energy strong electronic intervalley transfer into X and L valleys of

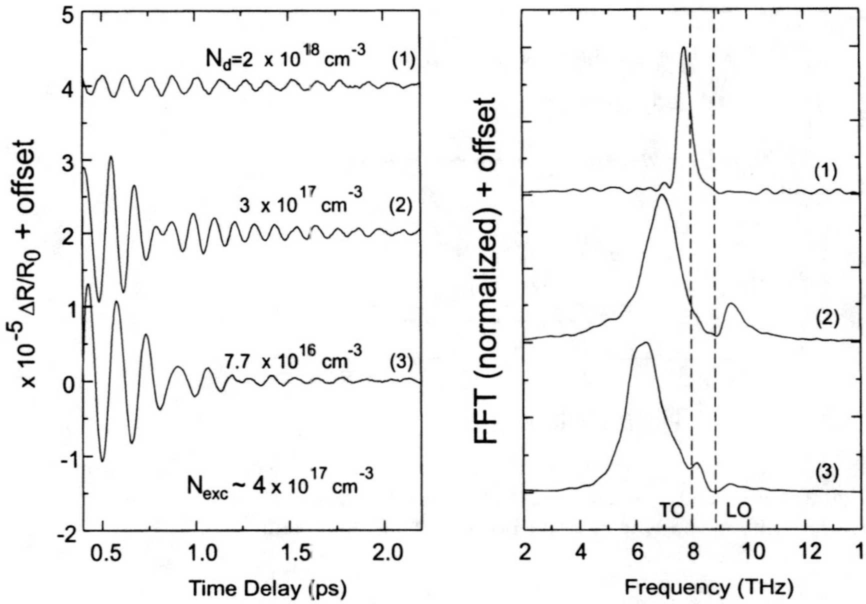

Fig. 4.12. Oscillatory part of isotropic reflectivity changes of differently doped GaAs samples at a constant excitation density (left) and Fourier transforms of the data (right, from [4.57])

the conduction band within the first 100 fs has been reported [4.55]. This leads in turn to a multicomponent plasma not having a well-defined frequency.

By using a Ti:sapphire laser, both lateral and k-space inhomogeneities can be suppressed. Figure 4.11 depicts coherent plasmon–phonon oscillations detected in the electro-optic reflectivity change at 1.5 eV. The crystal is (100)-oriented n-doped GaAs with a doping density of $N_\mathrm{D} = 3 \times 10^{17}\mathrm{cm}^{-3}$. The data are recorded using a 5 times larger pump- than probe-spot diameter. Thereby it is possible to scan the probe spot across the pump spot, thus detecting different optically excited densities. The maximum density in the center of the pump spot is $1 \times 10^{18}\mathrm{cm}^{-3}$. At this point a rapidly dephasing mode with a frequency close to the screened LO phonon is observed. For detection spots further away from the center, the frequency of the fast dephasing component decreases according to the dispersion of the lower plasmon branch and the unscreened LO phonon dominates the reflectivity changes. These results demonstrate that the mode-beating observed at high densities using 2 eV excitation (Fig. 4.10) stems from different contributions within the laterally inhomogeneous plasma.

After having demonstrated the time-resolved detection of coherent plasmon–phonon modes we investigate the effect of a cold carrier plasma introduced via doping of the sample. This investigation is expected to shed light on the mechanism relevant for the dephasing of coherent plasmon oscil-

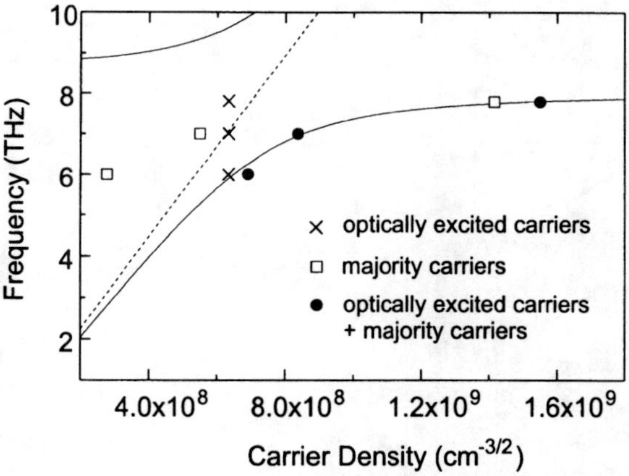

Fig. 4.13. Plasmon–phonon dispersion (solid line), plasmon frequency (dashed line) and the frequencies obtained from the time-resolved data as a function of the optically excited density (crosses), the doping densities of the sample (squares) and the sum of both (filled circles) (from [4.57])

lations [4.57]. Figure 4.12 shows coherent plasmon–phonon signatures from n-doped GaAs crystals with different doping densities at a constant optical excitation density. Figure 4.13 depicts the frequencies derived from the experimental data in comparison with the theoretical plasmon–phonon dispersion curve. Clearly, the data only follow the expected dependence if the sum over the optically excited density *and* the cold plasma resulting from the doping is taken into consideration. This observation gives strong evidence that the initially thermalized background plasma participates in the coherent plasmon oscillation. A closer analysis of the dephasing time of the plasmon oscillations reveals that it decreases if the relative contribution of the optically excited density increases, even when the total carrier density is kept at a constant level. This is a clear indication of the importance of electron–hole scattering for the plasmon dephasing [4.57].

Recently, terahertz radiation from coherent plasmon modes has been reported by *Kersting* et al. [4.58]. In time-resolved reflectivity experiments on GaAs at 77 K, plasmons far away from the phonon resonance were detected by *Sha* et al. [4.59]. Furthermore, coherent plasmon–phonon modes as presented here were observed in time-resolved Second Harmonic Generation (SHG) from bulk GaAs [4.60]. This detection is closely related to the electro-optic detection of plasmon–phonon modes presented here. Due to the possible selectivity of SHG towards the surface contribution, this technique opens an intriguing way to study the temporal dynamics of interface phonons [4.61, 4.62].

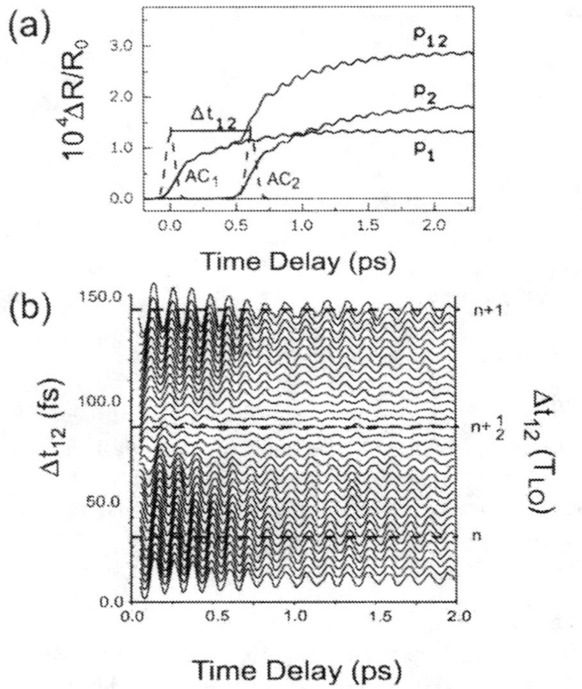

Fig. 4.14. (**a**) Electro-optic reflectivity changes of a (100) GaAs surface covered with an indium–tin-oxide Schottky contact on top. Two fs laser pulses at 2 eV separated by the delay Δt_{12} are used for coherent-phonon generation; p_1 and p_2 denote the data obtained with single-pulse excitation, p_{12} the signal for double-pulse excitation. (**b**) The extracted oscillatory contribution of the signal after the second pump pulse for different pulse separations Δt_{12} in units of the LO-phonon period ($n = 4$)

4.4.3 Coherent Control of LO Phonons

The phase information obtained from the time-resolved detection of the coherent amplitude opens the way towards the coherent control of that amplitude. Multiple successive pulses have been applied for a selective enhancement of certain vibrational modes in molecular systems [4.63]. Especially the recent advances in the shaping of femtosecond pulses [4.64] opens the way to driving the phonon dynamics in a well-defined way. By this technique, lattice distortions can be achieved that could not be accomplished with a single intense optical pulse due to the saturation of the optical transition.

The concept of coherent control over the phonon dynamics is readily illustrated by means of a two-pulse excitation experiment [4.65]. Figure 4.14 depicts the transient electro-optic reflectivity changes obtained on a (100) GaAs surface. The excitation of coherent LO phonons is achieved via the field-screening mechanism. Two successive pump pulses impinge upon the sample. Their intensity and time delay is adjusted in such a way that the second pulse provides a driving force for a coherent amplitude equivalent to the amplitude persistent from the first pulse. In addition, the initial surface field is adjusted via a transparent Schottky contact, in order not to screen the surface field completely already with the first pulse. By carefully adjusting the pulse separation, the driving force is in-phase or out-of-phase with the primarily generated coherent mode. In this way, complete destruction or resonant enhancement of coherent LO phonons is observed. This method enables the generation of coherent LO phonons for a well-defined time interval shorter than the intrinsic dephasing time of LO phonons. Similar experiments are performed in Sb, where the A_1-mode is manipulated in a similar way [4.37]. In Bi–Sb mixed crystals, the Bi–Bi, Bi–Sb, and Sb–Sb vibrations could be enhanced and canceled by applying femtosecond pulse trains [4.66]. Intriguing experiments have been reported on the coherent control of the dynamics of degenerate phonon modes, which allow even the generation of circularly polarized phonons [4.67]. Recently, the coherent control of acoustic phonons in superlattices was accomplished, where acoustic backfolded modes of different order were silenced out or enhanced by applying multiple successive pump pulses. By this method a high-sensitivity detection of higher-order modes can be achieved [4.68] (for the generation of acoustic phonons see Sect. 4.5.3).

4.5 Coherent Phonons in Low-dimensional Semiconductors

4.5.1 Coupled Intersubband–Plasmon Phonon Modes in Quantum Wells

Carrier–phonon interaction is one of the most important processes leading to energy relaxation in semiconductor devices, e.g., in heterostructure lasers [4.69]. The relaxation channels are strongly modified compared to the bulk semiconductor due to the introduction of discrete energy levels. The relaxation rates may strongly depend on whether the level spacing is smaller or larger than an LO-phonon energy [4.70]. Especially at the high carrier densities present in semiconductor lasers during operation, new collective modes can be formed based on the coupling between intersubband transitions and phonons. Such modes have been denoted as intersubband-plasmon modes; they have been observed in Raman scattering experiments in two-dimensional electronic systems [4.71, 4.72, 4.73]. A review of the investigations of elementary excitations in these systems by inelastic light scattering has been given by *Pinczuk* and *Abstreiter* [4.74].

In the case of femtosecond excitation a non-equilibrium population of several subbands can be achieved. This non-equilibrium population alters the dielectric response of the system along the growth direction. If the associated change in the polarization is fast enough on the time-scale of the transition

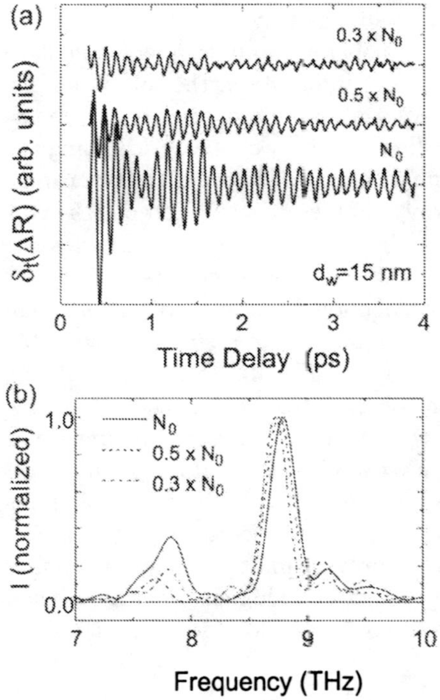

Fig. 4.15. (**a**) Oscillatory traces of electro-optic reflectivity changes of a 15 nm wide MQW sample at different excitation densities. The experiments are performed with a laser energy of 1.5 eV at 300 K. The maximum density N_0 corresponds to a bulk density of $3\times10^{18}\text{cm}^{-3}$. (**b**) Fourier transforms of the data in (**a**) normalized to the peak close to the LO phonon

frequencies, the new resonances of the dielectric response are excited coherently and can be probed via the electro-optic detection scheme described in Sect. 4.4.1. We would like to note that effects based on field screening are strongly altered in 2D due to the confinement of the carriers along the growth direction of the heterostructures.

The dielectric response of the two dimensional system, considering several subbands, may be written as [4.75]:

$$\epsilon(\omega) = \epsilon_\infty \left(\frac{\omega_{\text{LO}}^2 - \omega^2}{\omega_{\text{TO}}^2 - \omega^2} \right) + 4\pi \sum_{\alpha,i,j} \frac{|d_{ij}^\alpha|^2 (n_i^\alpha - n_j^\alpha)}{\varepsilon_i^\alpha - \varepsilon_j^\alpha + \omega}, \tag{4.12}$$

where α denotes the heavy-hole and conduction band, i and j are the subband indices, $\epsilon_{i,j}$ are the subband energies, d_{ij}^α is the dipole matrix element between level i and j, and $n_{i,j}^\alpha$ is the population of the subbands summed over perpendicular momenta. We note that the second part of (4.12) gives a contribution to the dielectric function if $n_i^\alpha \neq n_j^\alpha$, i.e., two adjacent levels are populated differently. The electronic dielectric function has been calculated for different quantum well widths and under the assumption of a thermal population of the subbands [4.75].

Figure 4.15 depicts experimental results for 15 nm wide multiple quantum wells excited resonantly at the lowest interband transitions for three different excitation densities. The highest excitation density corresponds to

$3\times10^{18}\text{cm}^{-3}$ in bulk GaAs. At this level the bulk coherent-phonon spectra are fully dominated by the screened LO phonon at the TO frequency (see Sect. 4.4.2). In the quantum well, the screening is strongly suppressed, since the carriers cannot move freely along the (100) direction. With increasing excitation density in the quantum well a mode beneath but close to the TO frequency evolves, while the mode close to the LO phonon slightly shifts to higher energies. Both features can be qualitatively explained by the shift of the resonances introduced by a non-thermal population of the different subbands [4.75].

Further time-resolved experiments on plasmon–phonon coupling have been reported by Baumberg and Williams for a GaAs/$Al_xGa_{1-x}As$ 2D electron gas [4.76]. An increase in the screened LO phonon at TO frequency was observed for increased optical excess energy above the GaAs bandgap; it is induced by an increase in the excitation density.

4.5.2 Coupled Coherent Bloch-Phonon Oscillations in Superlattices

The previous sections discussed carrier–phonon interaction studied via coherent phonons at high excitation densities. An intriguing situation arises in quantum confined systems, when the electronic level spacing equals the LO-phonon energy. In this case a large transition matrix element for phonon-assisted relaxation is anticipated. The electronic level spacing in heterostructures can be manipulated by means of electric or magnetic fields. Inelastic light-scattering experiments have been performed on Landau levels separated by an LO-phonon energy, leading to the observation of magneto-polarons [4.77]. In semiconductor superlattices, the energy separation of Wannier–Stark levels can be tuned via an applied electric field over a range given by the electronic miniband width [4.78]. In this system, double and triple resonant Raman scattering experiments have been performed revealing increased scattering cross sections under resonance conditions [4.79, 4.80].

Semiconductor superlattices open the unique possibility to coherently excite electronic wavepackets with tunable oscillation frequencies, i.e., Bloch oscillations, which are accessible by several experimental techniques [4.81, 4.82, 4.83, 4.84, 4.85]. The Bloch frequency is determined by $\nu_{\text{BO}} = eFd/h$, where d is the superlattice period and F the applied electric field. The Bloch wavepackets are associated with a macroscopic polarization oscillating along the growth direction of the superlattice, leading to the emission of THz radiation [4.83]. For details of Bloch oscillations we refer to the review by *Kurz* et al. [4.85].

Here we show results on coherent Bloch oscillations GaAs/$Al_xGa_{1-x}As$ superlattices with large miniband widths, i.e., minibands with energy widths equal or larger than the LO-phonon energy, where the oscillation frequency can be tuned into resonance with the GaAs LO phonon. The detection is based on an optical anisotropy induced by the coherent electronic polarization

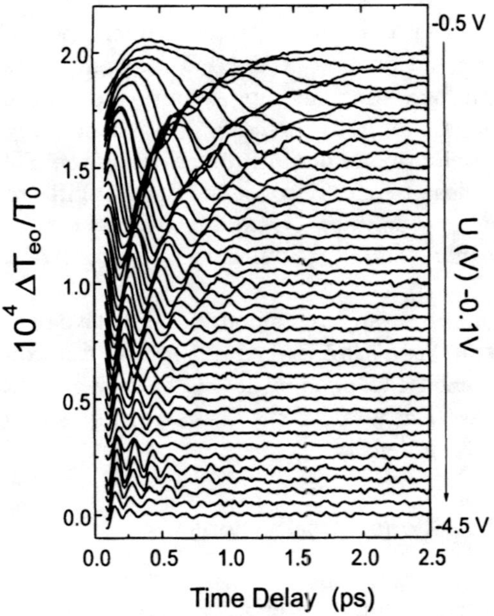

Fig. 4.16. Coherent Bloch oscillations extracted from anisotropic transmission changes in a 35 period $GaAs/Al_{0.3}Ga_{0.7}As$ superlattice of 67 Å well width and 17 Å barrier width. The calculated first electronic miniband width of the superlattice is 36 meV. The lattice temperature is 10 K and the excitation density is $3\times10^9 cm^{-2}$. The data are depicted for different voltages applied to the superlattice that was embedded into a Schottky diode [4.31]

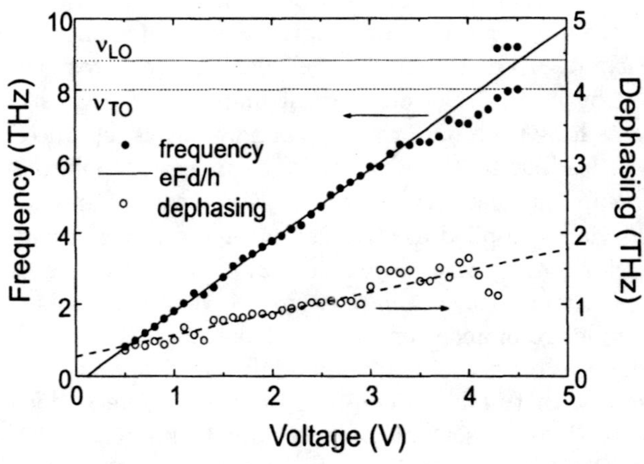

Fig. 4.17. Frequency and dephasing rate of Bloch oscillations depicted in the previous picture. The linear slope of the frequency obeys the Bloch relation $\nu = eFd/h$ [4.31]

[4.84], which is accomplished in basically the same experimental set up as the electro-optic detection of coherent LO phonons in GaAs, but in transmission geometry.

Figure 4.16 depicts the extracted Bloch oscillations detected at 10 K for different voltages applied to the sample. The excitation density corresponds to 3×10^9 electron–hole pairs per cm^2, which is more than three orders of magnitude lower than the densities used for the excitation of coherent LO phonons in GaAs via surface-field screening. An increase in frequency can be clearly observed, from some 100 GHz close to 10 THz, which is approximately the limit given by the miniband width. Beside the rise in frequency, a decrease in oscillation amplitude is observed which is due to the increased Stark localization of the wavepackets with increasing field [4.86]. The effect of the reduced localization length, and the associated decreased Bloch amplitude, has been studied recently in terahertz emission [4.87] and more precisely in four-wave mixing experiments [4.88].

The change of the oscillation frequency can be derived from the Fourier transforms of the time-resolved data in Fig. 4.16 and is depicted in Fig. 4.17. The internal electric field in the sample is proportional to the applied voltage plus an offset voltage. The frequency follows over a wide range of voltages the linear relation expected for Bloch oscillations. Only for frequencies above 7 THz the frequencies start to deviate slightly from this relation. For higher voltages, a splitting into a lower and upper branch at the TO- and LO-phonon frequencies of GaAs, respectively, is observed. The origin of this splitting is presently unknown, since at these densities the LO phonon cannot be screened. Nonlinear interaction may account for the opening of a gap between the TO and LO phonon [4.89]. In addition to the deviations of the linear field–frequency relation, a decrease in the dephasing rate of about 30% is observed close to the phonon resonance (Fig. 4.17). This behavior is counterintuitive to an anticipated increased dephasing via resonant emission of LO phonons between Wannier–Stark levels, and would imply a stabilization of the electronic coherence via coupling to the lattice polarization. This behavior is in close analogy to the frequency dependence of the plasmon–phonon dephasing above resonance, when the lower-branch oscillation becomes more lattice like. It is important to note that this coupling observed in the superlattice occurs at two orders of magnitude lower carrier densities than necessary in bulk GaAs to tune the plasmon frequency into resonance with the LO phonon.

If the excitation density is increased by a factor of 3, we observe that the coherent electronic polarization launches coherent phonons in the superlattice. This excitation process is described by the last term in (4.4). Figure 4.18 depicts the oscillatory traces of Bloch-phonon oscillations in the same superlattice as investigated before. At reverse bias voltages below −3.8 V, the oscillations are purely electronic, with dephasing times in the sub-picosecond range. When the applied voltage is increased by −0.2 V, the fast dephasing Bloch oscillations transit into a long-living oscillation at the phonon frequency. The frequency of this oscillations matches the LO-phonon frequency of bulk GaAs. This observation demonstrates that via Bloch-phonon coupling coherent phonons can be excited at densities which are more than two orders of magnitude smaller than the densities necessary to drive coherent LO

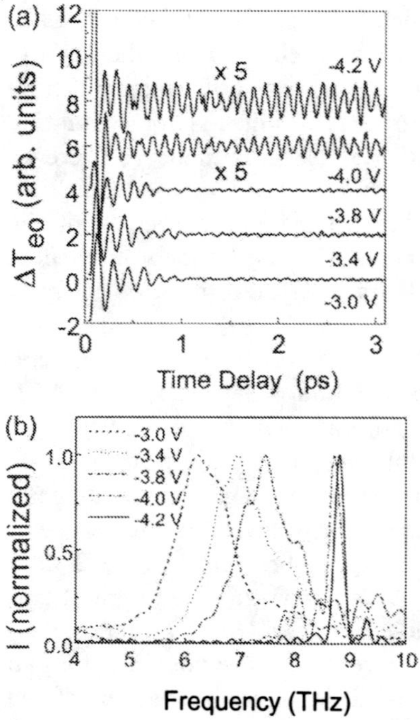

Fig. 4.18. (**a**) Time-resolved Bloch-phonon oscillations in a $GaAs/Al_{0.3}Ga_{0.7}As$ superlattice at 10 K for different applied reverse bias voltages. The data at the highest voltages are enlarged by a factor of 5. (**b**) Normalized Fourier spectra of the data. From [4.31]

phonons via field screening in bulk GaAs. This intriguing subject is under further experimental and theoretical investigation.

4.5.3 Coherent Acoustic Phonons in Superlattices

The understanding of low-frequency excitations in semiconductors is of paramount importance for a complete picture of scattering and dephasing processes. The detailed knowledge of the lifetimes of acoustic phonons is important for description of heat dissipation in semiconductor devices. The investigations discussed so far deal with the excitation of coherent optical phonons. Here we present recent developments in the observation of coherent acoustic phonons in GaAs/AlAs based heterostructures (for further developments in low-dimensional systems see also Sect. 4.8). The acoustic phonon branch is directly accessible by means of Brillouin scattering.

In semiconductor superlattices the effect of zone folding leads to a series of acoustic phonons with $\omega \neq 0$ for $\boldsymbol{k} = 0$ [4.90]. Zone-folded acoustic phonons have been intensively studied during the last years by Raman scattering [4.91, 4.92]. The first time-resolved observation of coherent acoustic phonons succeeded in GaAs/AlAs superlattices under resonant excitation of the first interband transitions [4.93]. The samples consist of 15 (18) monolayers GaAs wells and AlAs barriers. The coherent acoustic modes induce a

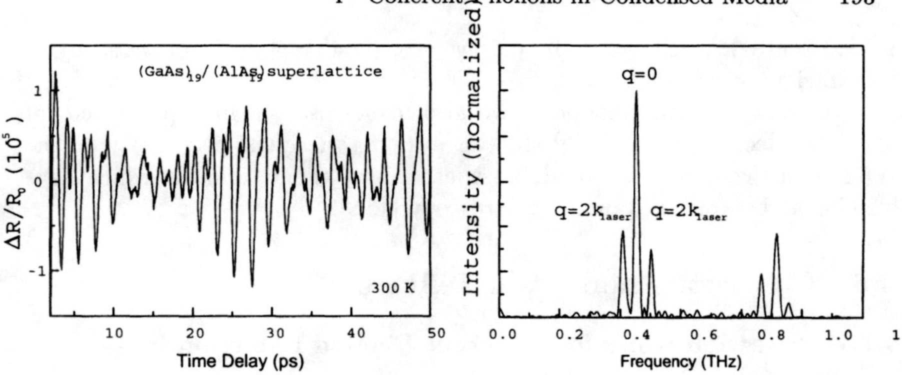

Fig. 4.19. Time-resolved reflectivity changes from a 19 monolayers GaAs/19 monolayers AlAs superlattice at room temperature (left). Fourier transform of the time-resolved data (right, adapted from [4.95])

change in the reflectivity, which is detected by the measurement of the time-derivative of the reflectivity change [4.94]. This technique can be applied since the reflectivity change induced by the electronic excitation is much larger than the phonon-induced oscillatory component. The data suggest that a stress in the wells is generated by the optical excitation, thus leading to the excitation of a mode of B_2 symmetry within the D_{2d} point group of the superlattice. However, this observation is in disagreement with Raman selection rules, i.e., the B_2 mode is not Raman active.

More recently we [4.95] and other groups [4.96, 4.97] investigated the generation of coherent acoustic phonons in GaAs/AlAs superlattices in more detail. Figure 4.19 shows high-resolution time-resolved reflectivity changes obtained by excitation of a 19 monolayers GaAs/19 monolayers AlAs superlattice in resonance with the first interband transition. The reflectivity changes exhibit a complicated oscillatory structure in the time domain which stems from the superposition of 6 different frequencies. The reflectivity change of the order of $10^{-3}\Delta R/R_0$, which is induced by the electronic excitation of the superlattice, has been subtracted. The observed modes belong to two triplets of the first and second-order backfolded acoustic phonon spectrum. Each triplet has a strong center mode which corresponds to the $\boldsymbol{k} = 0$ mode of A_1 symmetry. The two smaller peaks exactly match the calculated frequencies of phonons with a wavevector of $\boldsymbol{k} = 2\,\boldsymbol{k}_{\text{laser}}$. This observation points towards ISRS in forward ($\boldsymbol{k} = 0$) and backward scattering direction ($\boldsymbol{k} = 2\,\boldsymbol{k}_{\text{laser}}$) as the driving force [4.95]. The detection process is observed to be resonantly enhanced at the lowest interband transitions and is assumed to be based on the energy shift of the interband transitions coupled to the acoustic mode via the deformation potential. The dephasing times of the acoustic modes are in the range of 100 ps. The linewidths in the Fourier spectrum in Fig. 4.19 are still limited by the finite time window of the experiment. By taking time scans over time delays > 200 ps we recently obtained linewidths of 0.3 GHz

($< 0.01\ \mathrm{cm}^{-1}$) which opens the pathway for high-resolution spectroscopy in this intriguing material system.

Coherent acoustic phonons have also been observed in superlattices of the Fibonacci sequence [4.98]. In this system, the quasi-periodicity of the superlattice generates a self-similar spectrum of phonon modes, which have also been observed in Raman spectroscopy [4.99].

4.6 Coherent Phonons in Tellurium

4.6.1 Selection Rules for Coherent-Phonon Detection in Te

For the case of bulk GaAs we illustrated in Sect. 4.4, how different nonlinear contributions to the coherent-phonon detection process can be separated. Here we would like to show, for Te single crystals, that different phonon modes can be separated on the basis of the symmetry of the relevant Raman tensors. Tellurium is a perfect candidate for this purpose, since its crystal symmetry results in a set of only Raman-active, Raman- and ir-active, and only ir-active lattice modes. The phonon spectrum consists of 6 optical phonon modes. Te crystallizes in a hexagonal lattice (space group D_3^4 or D_3^6) containing three atoms per unit cell arranged in a helix along the c-axis. The lattice vibrations consist of a fully symmetrical, only Raman-active A_1-mode (3.6 THz), two degenerate Raman- and ir-active E-modes ($E'_{\mathrm{TO/LO}}$: $2.76/-3.09$ THz, $E''_{\mathrm{TO/LO}}$: $4.22/-4.26$ THz) and one only ir-active A_2-mode ($A_{2,\mathrm{TO/LO}}$: $2.6/-2.82$ THz) [4.100]. The internal polarization is either perpendicular (E-modes) or parallel (A_2-mode) to the c-axis. This will be relevant in the next section for the detection of terahertz emission.

The fully symmetric A_1-mode in Te is driven by the DECP mechanism, as has been demonstrated experimentally and confirmed theoretically [4.20]. However, the excitation of other modes of less symmetry cannot be explained within the DECP model, but is rather based on ISRS.

We applied to Te the same simultaneous detection of isotropic and anisotropic reflectivity changes as already introduced for the electro-optic detection of LO phonons in GaAs in Sect. 4.4.1. Again, the term *isotropic* is used to express that the phonon induced reflectivity changes do not depend on the relative angle between the probe pulse polarization and a certain crystal axis, while *anisotropic* means that two orthogonal components of the reflected probe pulse are subtracted from each other. This notation should be clearly distinguished from the Raman notation of polarized and depolarized spectra.

Figure 4.20 depicts the time-resolved isotropic and anisotropic reflectivity changes recorded in Te at a surface perpendicular to the c-axis. The experiments are performed with a CPM dye laser. The isotropic reflectivity shows a strong modulation at the frequency of the A_1-mode, while the anisotropic reflectivity exhibits a more complicated oscillatory behavior. The Fourier transformed time-domain data reveal the two E-modes and the absence of the A_1-mode in the anisotropic reflectivity change.

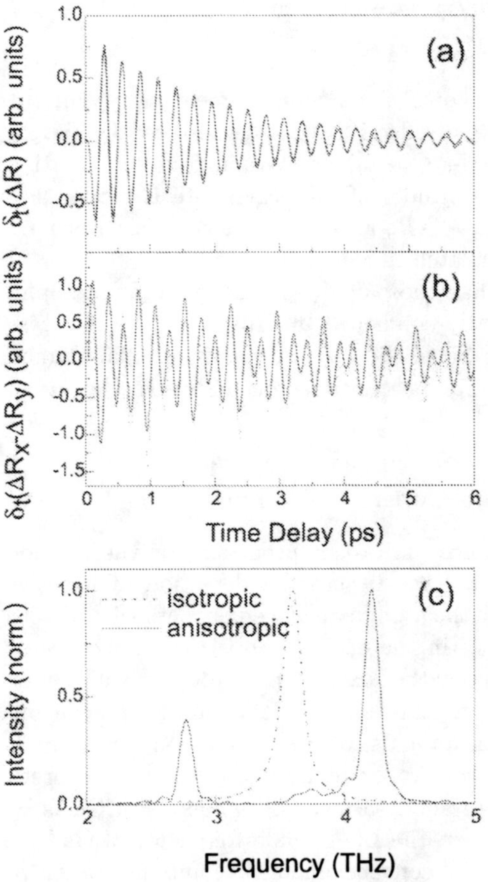

Fig. 4.20. Oscillatory component of the (**a**) isotropic and (**b**) anisotropic reflectivity change from single crystal Te, surface perpendicular to the c-axis, obtained at 2 eV. The data are recorded with a time-differential method [4.94]. Numerical fits (thin lines) are hard to distinguish from the experimental data. (**c**) Fourier transform of the time-resolved signals

The manifestation of phonon modes in the anisotropic reflectivity changes results from the off-diagonal elements of the Raman tensor. The anisotropic reflectivity can be expressed as

$$\Delta R_j - \Delta R_k \sim \boldsymbol{E}_j^{\mathrm{r}}(\partial\chi/\partial Q)\boldsymbol{E}^{\mathrm{i}}Q - \boldsymbol{E}_k^{\mathrm{r}}(\partial\chi/\partial Q)\boldsymbol{E}^{\mathrm{i}}Q, \tag{4.13}$$

where $\boldsymbol{E}^{\mathrm{i}}$ and $\boldsymbol{E}^{\mathrm{r}}$ are the incident and reflected probe fields, respectively. $(\partial\chi/\partial Q)$ denotes the Raman tensor, which contains the following non-zero elements for the two doubly degenerate E-modes [4.100]:

$$E(x) \quad : \quad (\partial\chi/\partial Q)_{xx} = -(\partial\chi/\partial Q)_{yy} = c\,,$$
$$(\partial\chi/\partial Q)_{yz} = (\partial\chi/\partial Q)_{zy} = d\,;$$

$$E(y) \quad : \quad (\partial\chi/\partial Q)_{xy} = (\partial\chi/\partial Q)_{yx} = -c \,,$$
$$(\partial\chi/\partial Q)_{xz} = (\partial\chi/\partial Q)_{zx} = -d \,.$$

The Raman tensor of the A_1-mode contains diagonal elements only. For the c$\perp$ surface, we obtain the anisotropic reflectivity change induced by the E-modes: $\Delta R_x - \Delta R_y \sim \sqrt{2}c(Q_{E(x)} + Q_{E(y)})$, i.e., a non-zero contribution from the coherent displacement amplitudes of the degenerate E-modes. For the fully symmetric A_1-mode, $\Delta R_x - \Delta R_y$ is zero. These considerations are confirmed by the experiments illustrated in Fig.4.20.

The detection of modes other than those of A_1 symmetry in (anisotropic) reflectivity changes has been recently confirmed by *Hase* et al. for the case of Bi [4.37] and *Garrett* et al. for Sb [4.36]. In transparent media, Raman selection rules in the phonon detection process have been observed by *Liu* et al. in $LaAlO_3$ [4.101].

4.6.2 Terahertz Emission from Coherent Phonons

In the previous section the relevance of Raman processes for the phonon detection has been established. Here we discuss the detection of ir-active but Raman-inactive phonons. Since the atomic displacements associated with these modes do not lead to changes in the optical properties in the visible range of the spectrum, they can be neither excited nor detected via Raman processes. By means of time-resolving the terahertz emission from semiconductors following pulsed excitation, an emission at the phonon frequency is expected. The terahertz emission from coherent phonons was first theoretically proposed for coherent (Raman and ir-active) LO phonons in GaAs by *Kuznetsov* and *Stanton* [4.9]. However, the LO-phonon frequency of GaAs is not within the detection bandwidth of conventional dipole antennas used for time-resolved terahertz detection. Therefore we have chosen to investigate the THz emission from single crystalline Te, which exhibits ir-active phonons in a lower frequency range.

In the THz-emission experiments, the sample is excited under 45° incidence by Ti:sapphire laser pulses with 1.75 eV photon energy and a pulse duration of 150 fs. The coherent THz radiation emitted in the direction of the reflected optical beam is collected with two paraboloidal mirrors and detected with a submillimeter dipole antenna that is gated by a second time-delayed laser pulse. The current in the dipole antenna, which is proportional to the incident electric field, is recorded as a function of the time delay [4.40]. The sensitivity of the detection system peaks at about 1 THz and extends up to 3 to 4 THz [4.40].

Figure 4.21a shows the measured electric field emitted from a surface perpendicular to the c-axis of tellurium. The data are compared to the emission spectrum of a broad-band emitter (InP), where the emission is based on the screening of the surface field [4.40]. The signal from Te consists of a strong initial emission followed by a periodically modulated tail. The strong

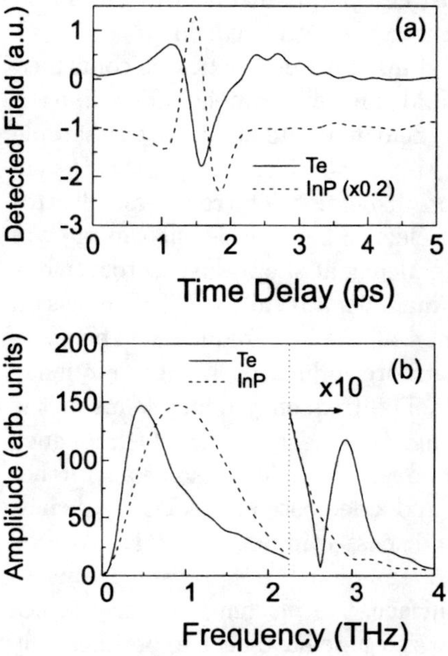

Fig. 4.21. (**a**) THz emission from Te (surface perpendicular to the *c*-axis) and InP. (**b**) Fourier transform of the time-domain data in (**a**). The high-frequency part is enlarged by a factor of 10. Adapted from [4.10]

peak is attributed to a polarization that results from the ultrafast build-up of an electric photo-*Dember* field [4.29], which is driven by the strong carrier gradient at the surface and the difference in electron and hole-diffusion coefficients [4.30]. This effect is amplified by differences in the transient electron and hole temperatures. Numerical simulations confirm a Dember field build up with amplitudes of 50 kV/cm and a rise-time of 100 fs. The screening of depletion-layer fields as the source of the radiation can be neglected in Te, because of a low density of charged surface states within the small bandgap [4.102]. The amplitude of the initial THz signal due to the Dember field is only a factor of 5 weaker than the emission from a polar semiconductor with strong surface fields (e.g., InP) for the same excitation power. Figure 4.21b depicts the amplitude of the Fourier transform of the time-domain signal. The frequency spectrum reveals, for Te, a broad peak at approximately 500 GHz. At high frequencies, the signal decreases to nearly zero at 2.62 THz before reaching a second maximum at about 2.9 THz. For clarity, this part of the spectrum is enlarged by a factor of 10. The high-frequency spectrum of a broadband emitting surface (InP, also $\times 10$, not normalized) representing the antenna response is shown for comparison. By polarization analysis of the high-frequency emission it can be shown that it stems from the coherently excited A_2-mode at 2.82 THz, which is only ir active. This mode has an internal polarization parallel to the c-axis, i.e., perpendicular to the excited surface. Thus this mode couples effectively to the Dember field.

The excitation mechanism via the build-up of a Dember field is confirmed by numerical calculations of the non-equilibrium electron–hole distribution close to the surface [4.30]. The Dember field perpendicular to the sample surface has a transverse wavevector due to the 45° incidence of the optical pulse. According to the phonon–polariton dispersion for transverse electromagnetic waves, the polarization *and* the electric field inside the sample will dominantly oscillate at the LO-phonon frequency at small wavevectors, while the polarization at the TO frequency is small. As a result, the THz emission is much stronger at the LO frequency than at the TO frequency. The spectral shape of the THz emission is furthermore influenced by the frequency dependence of the outcoupling efficiency. The frequency dependence of the refractive index makes the outcoupling most efficient at the LO frequency and least efficient at that of the TO phonons. This effect leads to a further increase in intensity at the LO frequency and a decrease at the TO frequency. In addition, plasmon–phonon coupling as discussed in Sect. 4.4.2, is expected to change the spectral shape of the THz emission [4.11]. However, due to the large carrier gradient close to the surface, the plasma frequency is not well-defined. Recently, the results summarized here have been experimentally confirmed by *Tani* et al. in Te, PbTe and CdTe [4.12].

4.6.3 Impulsive-Mode Softening of Phonons

One important aspect of phonon spectroscopy is the fact that the time-resolved detection of the phonon frequency provides information on the dynamics of phase transitions. The transition from a crystalline to an amorphous state, e.g., is accompanied by strong changes in the phonon spectrum. Information about the origin of optically induced phase transitions is essential for the optimization of material processing based on high-power femtosecond laser sources, as well for future phase-change optical recording materials. One specific form of phase transitions proposed for excitation with femtosecond laser pulses is non-thermal melting, where the crystal is destabilized by the optical excitation before energy transfer from the excited carriers to the lattice can take place. Several investigations deal with this problem. Most of them are based on time-resolved SHG, which vanishes when the crystal loses its symmetry. It has been reported, that this process occurs on time scales which are shorter than the typical energy-transfer times from the electronic system to the lattice [4.103, 4.104, 4.105]. Theoretically it has been predicted that a non-thermal phase transition may occur when more than 9% of all valence electrons are excited into antibonding conduction-band states [4.106, 4.107].

For an investigation of this intriguing phenomenon, we investigate the coherent-phonon dynamics in single-crystal Te excited by amplified CPM laser pulses of 2 eV photon energy [4.108]. This laser system delivers pulse

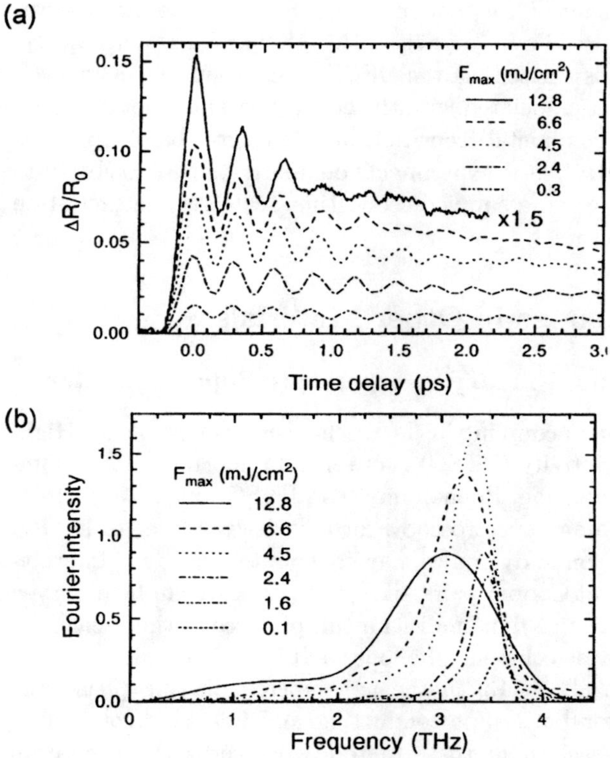

Fig. 4.22. (**a**) Time-resolved reflectivity changes from single-crystal Te for different excitation fluences as given in the figure. (**b**) Non-normalized Fourier transforms of the data in (**a**) (adapted from [4.108])

energies of 1.2 μJ at a repetition rate of 6.8 kHz. Figure 4.22 shows the time-resolved reflectivity changes at different pump fluences. The reflectivity changes are dominated by the A_1-mode excited via DECP as has already been shown in Sect. 4.6.1. The oscillation frequency decreases linearly for this range of fluences from 3.6 THz, i.e., the value of the unperturbed phonon resonance, to approximately 3 THz. A fit to the data shows that this mode softening is instantaneous. This is a clear hint for non-thermal melting of the crystal, since the energy transfer to the lattice is expected to take place on a longer time scale. A numerical simulation of the phonon dynamics, including the effect of carrier diffusion on this time scale, gives an estimate for the critical excitation density necessary for the non-thermal melting of the crystal, which is about 20 % of all valence-band electrons [4.109]. A careful comparison of the reflectivity change induced by the coherent phonons with the reflectivity changes induced by heating the crystal allows one to determine the relative lattice displacement of 0.36 % of the equilibrium lattice constant for the highest excitation density in Fig. 4.22 [4.110].

In Ti_2O_3 a transient shift of the phonon frequency, similar to the observations in Te, has been observed by *Cheng* et al. [4.111]. The Raman-active A_{1g} mode of Ti_2O_3 at 7 THz is excited via the DECP mechanism. An estimation of the lattice displacement associated with the coherent atomic motion shows that the semiconductor–semimetal transition in this narrow-bandgap material is modulated at 7 THz. These experiments demonstrate the feasibility to control material properties on a subpicosecond time scale via the excitation of large vibrational amplitudes.

4.7 Coherent Phonons in Other Materials

4.7.1 Coherent Phonons in High-temperature Superconductors

The role of electron–phonon coupling in the mechanism responsible for High-Temperature Superconductivity (HTSC) is of central importance for the understanding of this phenomenon. Phonon modes in HTSC's have been extensively investigated by Raman spectroscopy, for a review we refer to [4.112]. The study of coherent-phonon dynamics may contribute important information, since the dynamics of Cooper pairs and quasi particles can be observed concomitantly with the lattice dynamics on a sub-picosecond time scale.

The first observation of coherent phonons in HTSC materials has been reported by *Chwalek* et al. [4.113] in the non-superconducting phase and *Albrecht* et al. [4.114] for both superconducting and non-superconducting phases. As intriguing question in these materials concerns the excitation mechanism for coherent phonons – especially in the superconducting phase – and the relation to the optically induced breaking of Cooper pairs.

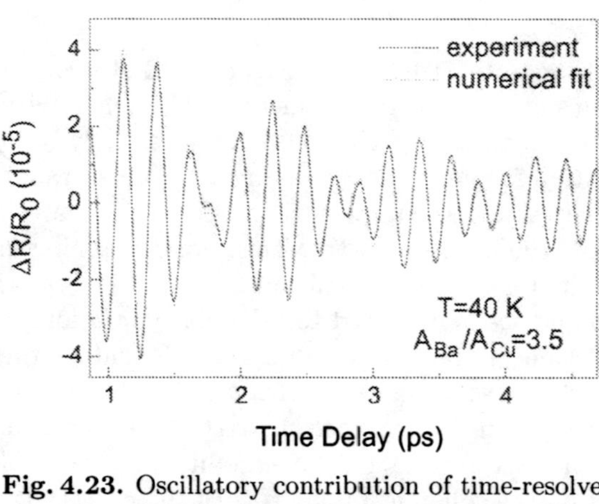

Fig. 4.23. Oscillatory contribution of time-resolved reflectivity change (solid line) of a $YBa_2Cu_3O_{7-\delta}$ reflectivity change thin film. The experiments are performed with a Ti:sapphire laser at 1.55 eV. The numerical fit is based on two sine functions with the Ba-mode frequency (3.6 THz) and the Cu-mode frequency (4.5 THz). From [4.115]

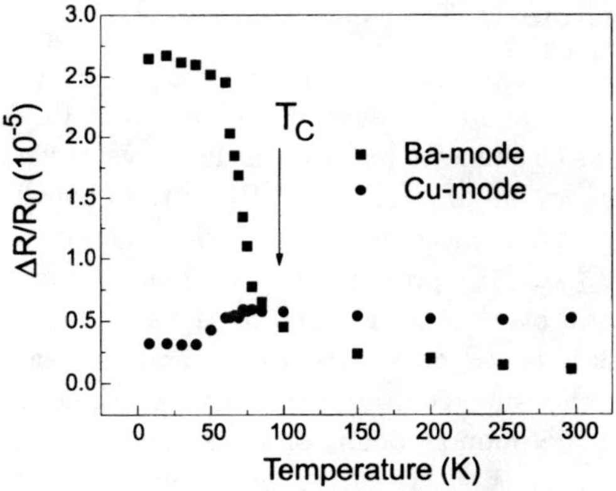

Fig. 4.24. Coherent-phonon amplitude of the Ba and Cu(2) modes in $YBa_2Cu_3O_{7-\delta}$ as a function of lattice temperature. From [4.115]

Figure 4.23 depicts the oscillatory contributions to time resolved reflectivity changes of $YBa_2Cu_3O_{7-\delta}$ (YBCO) in the superconducting phase [4.115]. The oscillations consist of a superposition of the A_{1g} modes of Ba and Cu(2) in the CuO_2 plane, with frequencies of 3.6 THz and 4.5 THz, respectively [4.116]. The amplitudes, phases and dephasing times are obtained with high accuracy from these fits. Figure 4.24 depicts the coherent-phonon amplitudes obtained from time-resolved reflectivity changes in YBCO measured at different lattice temperatures. While there is a weak temperature dependence of the amplitude of the coherently excited A_{1g}-mode of Ba and Cu(2) above T_c, the amplitude of the Ba mode strongly increases below T_c. In addition the phase of this mode changes strongly as a function of temperature, indicating a different excitation mechanism above and below T_c to be responsible for the coherent-phonon generation. The amplitude dependence is well reproduced by a two-fluid model below T_c [4.114,4.115], indicating a strong correlation between the coherent-phonon amplitude and the perturbation of the electronic system. Above T_c both modes are excited via DECP, leading to an impulsive bond-weakening of the lattice. Below T_c the excitation mechanism for the Ba mode can be explained by the ultrafast breaking of Cooper pairs induced by the optical pulse. The pair breaking strongly changes the conductivity in the CuO_2 planes. This produces a sudden change in the local crystal field of the Ba atoms, leading to their coherent displacement. The Cu(2) mode is less sensitive to the local crystal field changes. Similar changes of intensities of the Ba and Cu(2) modes observed in the time-resolved phonon amplitudes have been also observed in the ordinary Raman spectra [4.117]. The relevance of the sudden screening of the local crystal field and the implications on the generation of coherent phonons in HTSC have been confirmed recently in a systematic study on differently doped YBCO and in $Bi_2Sr_2CaCu_2O_8$ [4.115].

4.7.2 Coherent Phonon–Polaritons in Ferroelectric Crystals

In Sect. (4.6.2) we discussed the coherent excitation of an exclusively ir-active phonon mode via an electric Dember field and the associated emission of THz radiation. There are more direct ways to generate fir light inside a crystal via ISRS [4.118, 4.119], difference-frequency mixing [4.120, 4.121] or the optical Cerenkov effect [4.122]. In the generation mechanism based on difference-frequency mixing two laser fields E_k and E_l provided by two femtosecond laser pulses with spatial and temporal overlap are focused on the same spot on the sample [4.121]. This process is based on the second-order nonlinear susceptibility $\chi^{(2)}$ of the crystal. By choosing a certain angle between the two laser pulses a well-defined mode on the phonon–polariton dispersion can be selected corresponding to the wavevector $\boldsymbol{k} = \boldsymbol{k}_k - \boldsymbol{k}_l$. This method creates two counterpropagating polaritons within the laser spot. The generated coherent phonon–polariton can be probed via a third time-delayed laser pulse positioned closely on either side of the excitation spot. The polariton may be detected by probe light diffracted via the electro-optic effect. Since the difference-frequency mixing is performed with laser energies well below interband resonances, high peak intensities are necessary for these experiments. Therefore the experiments were performed with amplified laser pulses, e.g., with an amplified CPM laser at a photon energy of 2 eV, 60 fs pulse duration and a pulse energy of $\sim$ µJ.

By this technique, *Bakker* et al. were able to map out the phonon–polariton dispersion with high sensitivity in the low-frequency range in several ferroelectric crystals ($LiTaO_3$ [4.121] and $LiNbO_3$ [4.123]). Due to the high sensitivity of this method, new resonances in the dispersion could be detected at 0.95 THz in $LiTaO_3$ and at 1.3 THz, 2.4 THz, 3.4 THz and 4.1 THz in $LiNbO_3$. These frequencies, and the corresponding temperature dependence of the dephasing times, allowed the authors to develop a quantum-mechanical model for the ferroelectric phase transition. The newly observed resonances turned out to be quantum-mechanical tunneling resonances of the lowest phonon mode. A recent review on this subject can be found in [4.124].

Furthermore, time-resolved phonon–polariton studies have been performed by *Vallée* and *Flytzanis* [4.125], where phonon–polariton packets are generated in one crystal and the detection is accomplished in a second crystal separated by an air gap from the first. This technique allows the study of the influence of interfaces and surfaces on the phonon–polariton propagation.

4.8 Recent Developments

Recently there have been several interesting reports on the observation of coherent phonons in a variety of materials, which we would like to summarize here.

The application of laser pulses of 10 fs to 20 fs duration directly derived from a resonator enables one to time-resolve phonon oscillations in the range beyond 10 THz. *Fleischer* et al. recently demonstrated the observation of a coherently excited $A_g(1)$ mode at 14.9 THz in doped C_{60} molecules [4.126]. In addition to this mode, a low-frequency mode around 4.5 THz was observed which could not be observed in cw Raman experiments. These modes are attributed to the dopant atoms.

Hase et al. recently observed coherent plasmon–phonon modes also from the upper branch of the coupled-mode dispersion by applying 20 fs laser pulses [4.127]. These findings confirm that the frequency of coherent modes is given by both the optically excited carriers and the carriers from the background doping of the sample [4.57].

Coherent optical and acoustic modes have been observed recently in semiconductor quantum dots [4.128,4.129,4.130,4.131,4.132]. Resonant excitation of quantum dots in conjunction with strong electron–phonon coupling leads to a modulation of a four-wave-mixing signal with the frequency of the LO phonon in CdSe quantum dots [4.128, 4.129] and of the LO and TO mode in InP quantum dots [4.131]. Coherent acoustic modes could be observed in resonant pump–probe experiments on PbS quantum dots [4.132, 4.133]. The generation mechanism is based on deformation-potential coupling of the lowest-order spheroidal acoustic mode to the quantum-dot excitons [4.132]. The dominant damping mechanism has been identified as radiative loss from the quantum dots to the surrounding glass [4.133]. In contrast to the work on CdSe and InP quantum dots, the optical mode could not be observed in PbTe quantum dots, suggesting differences in the exciton–phonon coupling strength in quantum dots of IV–VI, II–VI and III–V composition.

A basic concept of quantum mechanics is Heisenberg's uncertainty principle for two observables represented by non-commuting operators. However, the quantum fluctuations of one observable may be reduced to below the vacuum limit at the expenses of the other such that the product of the fluctuations still obeys the uncertainty relation. A state prepared in this way is denoted as *vacuum squeezed state* and was first observed by *Slusher* et al. for photons [4.134]. Coherent phonons and polaritons may provide access to the realization of squeezed states in solids [4.135, 4.136]. Recently, *Garrett* et al. achieved the observation of phonon squeezed states in $KTaO_3$ [4.35]. The excitation is based on a second-order process leading to an impulsive change of the phonon frequency of transverse acoustic modes. The squeezing factor derived from the observation of coherently excited TA modes is in the range of 10^{-6}, which is still small compared to values obtained in squeezing experiments with photons. These findings on phonon squeezed states in solids are supported by experiments in $SrTiO_3$ by the same group [4.137].

The coherent control of electron–LO-phonon *scattering* has been achieved in femtosecond four-wave-mixing experiments, where the delay of two excitation pulses has been varied on a time scale of several tens of femtoseconds before electron-LO-phonon scattering is completed [4.138]. These experiments allowed the control of the scattering dynamics in a coherent fashion, thus leading to the suppression or enhancement of the scattering process within the time in which the process is still to be fully completed.

4.9 Conclusions

In this chapter we have presented an overview of the feasibilities of time-resolved detection of coherently excited lattice vibrations in the THz frequency range. In contrast to conventional Raman scattering, the generation and detection processes of coherent phonons can be well separated leading to interesting insights into the nonlinear interaction between sub-picosecond light pulses and semiconductors, superconductors and ferroelectric crystals, etc. In recent years, coherent-phonon spectroscopy has revealed many results important for the understanding of lattice dynamics and nonlinear phonon–photon interaction. This development was strongly driven by the improvement of sub-picosecond laser sources. Most recent developments of stable laser sources with pulse durations in the range of 10 fs and high peak powers up to μJ per pulse, in combination with pulse-shaping techniques, should allow us to perform a huge variety of intriguing experiments on coherent-phonon dynamics, including the coherent control over the lattice motion up to extremely large displacements. In comparison to cw inelastic light-scattering experiments, the field of time-resolved coherent-phonon spectroscopy is relatively young. The spread and availability of stable femtosecond lasers in all ranges of wavelengths and energies will certainly intensify the research in this field.

Acknowledgments

The authors thank the following people for their valuable contributions to this work: H. J. Bakker, A. Bartels, G. Bauer, M. Cardona, A. Förster, A. W. Ghosh, J. Geurts, P. Grosse, S. Hunsche, C. Jaekel, A. M. T. Kim, K. Köhler, W. Kütt, A. V. Kuznetsov, K. Mizoguchi, S. Nakashima, M. Nakayama, T. Pfeifer, H. G. Roskos, R. Scholz, and V. Wagner. We gratefully acknowledge support by the Deutsche Forschungsgemeinschaft and the Volkswagen Stiftung.

References

4.1 J. Shah: *Ultrafast Spectroscopy of Semiconductors and Semiconductor Nanostructure*, Solid State Sci. **115** (Springer, Berlin, Heidelberg 1996)
4.2 W. E. Bron, J. Kuhl, B. K. Rhee: Phys. Rev. B **34**, 6961 (1986)
4.3 D. von der Linde: In *Ultrashort Laser Pulses*, ed. by W. Kaiser, Topics Appl. Phys. **60** (Springer, Berlin 1988)
4.4 J. A. Kash, J. C. Tang: In *Spectroscopy of Nonequilibrium Electrons and Phonons*, ed. by C. V. Shank, B. P. Zakharchenya (Elsevier, Amsterdam, 1992) pp. 113–168

4.5 F. Vallée, F. Bogani: Phys. Rev. B **43**, 12049 (1991)
4.6 F. Vallée: Phys. Rev. B **49**, 2460 (1994)
4.7 F. Ganikhanov, F. Vallée: Phys. Rev. B **55**, 15614 (1997)
4.8 W. E. Bron: In *Coherent Optical Interactions in Semiconductors*, ed. by R. T. Phillips, NATO ASI Ser. B **330** (Plenum, New York, 1994) pp. 199–222
4.9 A. V. Kuznetsov, C. J. Stanton: Phys. Rev. Lett. **73**, 3243 (1994)
4.10 T. Dekorsy, H. Auer, C. Waschke, H. J. Bakker, H. G. Roskos, H. Kurz, V. Wagner, P. Grosse: Phys. Rev. Lett. **74**, 738 (1995)
4.11 A. V. Kuznetsov, C. J. Stanton: Phys. Rev. B **51**, 7555 (1995)
4.12 M. Tani, R. Fukasawa, H. Abe, S. Matsuura, K. Sakai, S. Nakashima: J. Appl. Phys. **83**, 2473 (1998)
4.13 G. C. Cho, W. Kütt, H. Kurz: Phys. Rev. Lett. **65**, 764 (1990)
4.14 G. C. Cho, H. J. Bakker, T. Dekorsy, H. Kurz: Phys. Rev. B **53**, 6904 (1996)
4.15 W. Kütt: In *Festkörperprobleme/Advances in Solid States Physics*, ed. by U. Rössler (Vieweg, Braunschweig 1992) Vol. 32, pp. 113–129
4.16 A. Lauberau, W. Kaiser: Rev. Mod. Phys. **50**, 607 (1978)
4.17 R. Merlin: Solid Stat. Commun. **102**, 207 (1997)
4.18 T. K. Cheng, S. D. Brorson, A. S. Kazeroonian, J. S. Moodera, G. Dresselhaus, M. S. Dresselhaus, E. P. Ippen: Appl. Phys. Lett. **57**, 1004 (1990)
4.19 T. K. Cheng, J. Vidal, M. J. Zeiger, G. Dresselhaus, M. S. Dresselhaus, E. P. Ippen: Appl. Phys. Lett. **59**, 1923 (1991)
4.20 H. J. Zeiger, J. Vidal, T. K. Cheng, E. P. Ippen, G. Dresselhaus, M. Dresselhaus: Phys. Rev. B **45**, 768 (1992)
4.21 Y.-X. Yan, E. B. Gamble, K. A. Nelson, J. Chem. Phys. **83**, 5391 (1985)
4.22 S. DeSilvestri, J. G. Fujimoto, E. B. Gamble, L. R. Williams, K. A. Nelson: Chem. Phys. Lett. **116**, 146 (1985)
4.23 Y.-X. Yan, K. A. Nelson: J. Chem. Phys. **86**, 6240 (1987)
4.24 Y.-X. Yan, K. A. Nelson: J. Chem. Phys. **86**, 6257 (1987)
4.25 J. Chesnoy, A. Mokthari: Phys. Rev. B **38**, 3566 (1988)
4.26 T. Dekorsy: Dissertation, RWTH Aachen (1997)
4.27 T. Pfeifer, T. Dekorsy, W. Kütt, H. Kurz: Appl. Phys. A **55**, 482 (1992)
4.28 T. Dekorsy, T. Pfeifer, W. Kütt, H. Kurz: Phys. Rev. B **47**, 3842 (1993)
4.29 H. Dember: Phys. Z. **32**, 554 (1931)
4.30 T. Dekorsy, H. Auer, H. J. Bakker, H. G. Roskos, H. Kurz: Phys. Rev. B **53**, 4005 (1996)
4.31 T. Dekorsy, A. M. T. Kim, G. C. Cho, K. Köhler, H. Kurz: In *Ultrafast Phenomena X*, ed. by P. F. Barbara, J. G. Fujimoto, W. H. Knox, W. Zinth (Springer, Berlin, Heidelberg 1996) pp. 382–384
4.32 B. B. Varga: Phys. Rev. **137**, 1896 (1965)
4.33 A. Mooradian, G. B. Wright: Phys. Rev. Lett. **16**, 999 (1966)
4.34 A. Mooradian, A. L. McWorther: Phys. Rev. Lett. **19**, 849 (1967)
4.35 G. A. Garrett, A. G. Rojo, A. K. Sood, J. F. Whitaker, R. Merlin: Science **275**, 1638 (1997)
4.36 G. A. Garrett, T. F. Albrecht, J. F. Whitaker, R. Merlin: Phys. Rev. Lett. **77**, 3661 (1996)
4.37 M. Hase, K. Mizoguchi, H. Harima, S. Nakashima, M. Tani, K. Sakai, M. Hangyo: Appl. Phys. Lett. **69**, 2474 (1996)
4.38 W. A. Kütt, W. Albrecht, H. Kurz: IEEE J. QE **28**, 2434 (1992)

4.39 M. Kuball, N. Esser, T. Ruf, C. Ulrich, M. Cardona, K. Eberl, A. Garcia-Cristobal, A. Cantarero: Phys. Rev. B **51**, 7353 (1995)
4.40 X.-C. Zhang, D. H. Auston: J. Appl. Phys. **71**, 326 (1992)
4.41 Q. Wu, X. C. Zhang: Appl. Phys. Lett. **70**, 1784 (1997)
4.42 Q. Wu, X. C. Zhang: In *Postdeadline Papers Conference on Lasers and Electro Optics* (Optical Society of America, Washington, DC 1997) pp. 1–3
4.43 A. Cantarero, C. Trallero-Giner, M. Cardona: Phys. Rev. B **39**, 8388 (1989)
4.44 A. Cantarero, C. Trallero-Giner, M. Cardona: Phys. Rev. B **40**, 12290 (1989)
4.45 M. J. Rosker, F. W. Wise, C. L. Tang: Phys. Rev. Lett. **57**, 321 (1986)
4.46 This system has become commercially available as a PC based system by GWU Lasertechnik, Germany
4.47 S. M. Sze: *Physics of Semiconductor Devices*, 2nd edn. (Wiley, New York 1981)
4.48 A. Yarif: *Introduction to Optical Electronics* (Holt, Rinehart and Winston, New York 1971)
4.49 L. Min, R. J. D. Miller: Appl. Phys. Lett. **56**, 524 (1990)
4.50 M. Hase, K. Mizoguchi, H. Harima, S. Nakashima, K. Sakai: Phys. Rev. B **58**, 5448 (1998)
4.51 J. G. Gay, J. D. Dow, E. Burstein, A. Pinzcuk: In *Light Scattering in Solids*, ed. by M. Balkanski (Flammarion, Paris 1971) pp. 33–38
4.52 R. M. Martin, L. M. Falicov: In *Light Scattering in Solids*, ed. by M. Cardona, Topics Appl. Phys. **8** (Springer, Berlin, Heidelberg 1975) pp. 79–145
4.53 M. Cardona: In *Light Scattering in Solids*, ed. by M. Cardona, Topics Appl. Phys. **50** (Springer, Berlin, Heidelberg 1982) pp. 19–178
4.54 T. Pfeifer, T. Dekorsy, W. Kütt, H. Kurz: In *Phonon Scattering in Condensed Matter VII*, ed. by M. Meissner, R. O. Pohl, Solid-State Sci. **112** (Springer, Berlin, Heidelberg 1993) p. 110
4.55 P. C. Becker, H. L. Fragnito, C. Brito-Cruz, J. Shah, R. L. Fork, J. E. Cunningham, J. E. Henry, C. V. Shank: Appl. Phys. Lett. **53**, 2089 (1988)
4.56 C. C. Cho: Dissertation, RWTH Aachen (1997)
4.57 G. C. Cho, T. Dekorsy, H. J. Bakker, R. Hövel, H. Kurz: Phys. Rev. Lett. **77**, 4062 (1996)
4.58 R. Kersting, K. Unterrainer, G. Strasser, H. Kaufmann, E. Gornik: Phys. Rev. Lett. **79**, 3038 (1997)
4.59 W. Sha, A. L. Smirl, W. F. Tseng: Phys. Rev. Lett. **74**, 4273 (1995)
4.60 W. de Jong, A. van Etteger, P. van Hall, T. Rasing: Surf. Sci. **377**, 355 (1997)
4.61 Y. M. Chang, L. Xu, H. W. K. Tom: In *Ultrafast Phenomena X*, ed. by P. F. Barabara, J. G. Fujimoto, W. H. Knox, W. Zinth, Springer Ser. Chem. Phys. **62**, (Springer, Berlin, Heidelberg 1996) pp. 391–392
4.62 Y. M. Chang, L. Xu, H. W. K. Tom: Phys. Rev. Lett. **78**, 4649 (1997)
4.63 A. M. Weiner, D. Leaird, K. A. Nelson: J. Opt. Soc. Am. B **8**, 1264 (1991)
4.64 A. M. Weiner: In *Coherence Phenomena in Atoms and Molecules in Laser Fields*, ed. by A. Bandrauk, S. C. Wallace (Plenum, New York 1992) pp. 277–289
4.65 T. Dekorsy, W. Kütt, T. Pfeifer, H. Kurz: Europhys. Lett. **23**, 223 (1993)
4.66 M. Hase, T. Itano, K. M. S. Nakashima: Jpn. J. Appl. Phys. **37**, L281 (1998)
4.67 M. M. Wefers, H. Kawashima, K. A. Nelson: J. Phys. Chem. Sol. **57**, 1425 (1996);M. M. Wefers, H. Kawashima, K. A. Nelson: J. Chem. Phys. **108**, 10248 (1998)

4.68 A. Bartels, T. Dekorsy, H. Kurz, K. Köhler: Appl. Phys. Lett. **72**, 2844 (1998)
4.69 J. Shah: In *Spectroscopy of Nonequilibrium Electrons and Phonons*, ed. by C. V. Shank, B. P. Zakharchenya (Elsevier, Amsterdam 1992)
4.70 S. Hunsche, K. Leo, H. Kurz, K. Köhler: Phys. Rev. B **50**, 5791 (1994)
4.71 E. Burstein, A. Pinczuk, D. L. Mills: Surf. Sci. **98**, 451 (1980)
4.72 D. Olego, A. Pinczuk, A. G. Gossard, W. Wiegmann: Phys. Rev. B **25**, 7867 (1982)
4.73 P. Brockmann, J. F. Young, P. Hawrylak, H. M. van Driel: Phys. Rev. B **48**, 11423 (1993)
4.74 A. Pinczuk, G. Abstreiter: In *Ligth Scattering in Solids*, ed. by M. Cardona, G. Güntherodt, Topics in Applied Physics **66** (Springer, Berlin, Heidelberg 1989) pp. 153–213
4.75 T. Dekorsy, A. M. T. Kim, G. C. Cho, H. Kurz, A. V. Kuznetsov, A. Förster: Phys. Rev. B **53**, 1531 (1996)
4.76 J. J. Baumberg, D. A. Williams: Phys. Rev. B **53**, R16140 (1996)
4.77 J. T. Devreese: *Polarons in Ionic Crystals and Polar Semiconductors* (North-Holland, Amsterdam, 1972)
4.78 F. Agulló-Rueda, J. Feldmann: In *Semiconductor Superlattices*, ed. by H. T. Grahn (World Scientific, Singapore 1995) pp. 99–154
4.79 F. Agulló-Rueda, E. E. Mendez, J. M. Hong: Phys. Rev. B **38**, 12720 (1988)
4.80 A. Alexandrou, M. Cardona, K. Ploog: Phys. Rev. B **38**, 2196 (1988)
4.81 K. Leo, P. Haring-Bolivar, F. Brüggemann, R. Schwedler, K. Köhler: Solid State Commun. **84**, 943 (1992)
4.82 J. Feldmann, K. Leo, J. Shah, D. A. B. Miller, J. E. Cunningham, S. Schmitt-Rink, T. Meier, G. von Plessen, A. Schulze, P. Thomas: Phys. Rev. B **46**, 7252 (1992)
4.83 C. Waschke, H. G. Roskos, R. Schwedler, K. Leo, H. Kurz, K. Köhler: Phys. Rev. Lett. **70**, 3319 (1993)
4.84 T. Dekorsy, P. Leisching, H. Kurz, K. Köhler: Phys. Rev. B **50**, 8106 (1994)
4.85 H. Kurz, H. Roskos, T. Dekorsy, K. Köhler: Proc. Royal Soc. Lond. A **354**, 2295 (1996)
4.86 T. Dekorsy, P. Leisching, W. Beck, R. Ott, Y. Dhaibi, H. Kurz, K. Köhler: Superl. Microstr. **15**, 11 (1994)
4.87 R. Martini, G. Klose, H. G. Roskos, H. Kurz, H. T. Grahn, R. Hey: Phys. Rev. B **54**, R14 (1996)
4.88 V. Lyssenko, G. Valusis, F. Loser, T. Hasche, K. Leo, M. Dignam, K. Köhler: Phys. Rev. Lett. **79**, 301 (1997)
4.89 A. W. Ghosh: private communication (1998)
4.90 M. Cardona, P. Yu: *Fundamentals of Semiconductors* (Springer, Berlin, Heidelberg 1996)
4.91 B. Jusserand, M. Cardona: In *Light Scattering in Solids V*, ed. by M. Cardona, G. Güntherodt: Topics Appl. Phys. **66** (Springer, Berlin, Heidelberg 1989) p. 49
4.92 T. Ruf: *Phonon Raman scattering in Semiconductors, Quantum Wells and Superlattices*, Springer Tracts in Modern Physics **142** (Springer, Berlin, Heidelberg 1998)
4.93 A. Yamamoto, T. Mishina, Y. Masumoto, M. Nakayama: Phys. Rev. Lett. **73**, 740 (1994)

4.94 T. Pfeifer, W. Kütt, H. Kurz, R. Scholz: Phys. Rev. Lett. **69**, 3248 (1992)
4.95 A. Bartels, T. Dekorsy, H. Kurz, K. Köhler: Phys. Rev. Lett. **82**, 1044 (1999)
4.96 K. Mizoguchi, K. Matsutani, M. Hase, S. Nakachima, M. Nakayama: Physica B **249–251**, 887 (1998)
4.97 T. Mishina, Y. Iwazaki, Y. Masumoto, M. Nakayama: Solid State Commun. **107**, 281 (1998)
4.98 K. Mizoguchi, K. Matsutani, S. Nakachima, T. Dekorsy, H. Kurz, M. Nakayama: Phys. Rev. B **55**, 9336 (1997)
4.99 R. Merlin: In *Ligth Scattering in Solids V*, ed. by M. Cardona, G. Güntherodt, Topics Appl. Phys. **66** (Springer, Berlin, 1989) pp. 214–232
4.100 P. Grosse, W. Richter: In *Landolt-Börnstein, Numerical Data and Functional Relationships in Science and Technology*, ed. by O. Madelung, New Series **17** (Springer, Berlin, Heidelberg 1983)
4.101 Y. Liu, A. Frenkel, G. A. Garrett, J. F. Whitaker, S. Fahy, R. Merlin: Phys. Rev. Lett. **75**, 334 (1995)
4.102 P. Grosse, M. Lutz, W. Richter: Solid State Commun. **5**, 99 (1967)
4.103 C. V. Shank, R. Yen, C. Hirlimann: Phys. Rev. Lett. **51**, 900 (1983)
4.104 H. W. K. Tom, G. D. Aumiller, C. H. Brito-Cruz: Phys. Rev. Lett. **60**, 1438 (1988)
4.105 P. Saeta, J. K. Wang, P. Siegal, N. Bloembergen, E. Mazur: Phys. Rev. Lett. **67**, 1023 (1991)
4.106 P. Stampfli, K. H. Bennemann: Phys. Rev. B **42**, 7163 (1990)
4.107 P. Stampfli, K. H. Bennemann: Phys. Rev. B **49**, 7299 (1994)
4.108 S. Hunsche, K. Wienecke, T. Dekorsy, H. Kurz: Phys. Rev. Lett. **75**, 1815 (1995)
4.109 S. Hunsche, K. Wienecke, T. Dekorsy, H. Kurz: In *Ultrafast Phenomena X*, ed. by P. F. Barabara, J. G. Fujimoto, W. H. Knox, W. Zinth, Springer Ser. Chem. Phys. **62** (Springer, Berlin, Heidelberg 1996) pp. 459–460
4.110 S. Hunsche, K. Wienecke, H. Kurz: Appl. Phys. A **62**, 499 (1996)
4.111 T. K. Cheng, L. H. Acioli, J. Vidal, H. J. Zeiger, M. S. Dresselhaus, E. P. Ippen: Appl. Phys. Lett. **62**, 1901 (1993)
4.112 C. Thomson: In *Light Scattering in Solids VI*, ed. by M. Cardona, G. Güntherodt, Topics Appl. Phys. **68** (Springer, Berlin, Heidelberg 1991) p. 285
4.113 J. M. Chwalek, C. Uher, J. F. Whitaker, G. A. Mourou, J. A. Agnostinelli: Appl. Phys. Lett. **58**, 980 (1991)
4.114 W. Albrecht, T. Kruse, H. Kurz: Phys. Rev. Lett. **69**, 1451 (1992)
4.115 C. Jaekel: Dissertation, RWTH Aachen (1997)
4.116 R. Liu, C. Thomson, W. Kress, M. Cardona, B. Gegenheimer, F. de Wette, J. Prade, A. D. Kulkarni, U. Schröder: Phys. Rev. B **37**, 7971 (1988)
4.117 B. Friedl, C. Thomsen, H. U. Habermeier, M. Cardona: Solid State Commun. **78**, 291 (1991)
4.118 T. P. Dougherty, G. P. Wiederrecht, K. A. Nelson: Ferroelectrics **120**, 79 (1991)
4.119 J. Etchepare, G. Grillon, J. C. Loulergue, A. Antonetti, M. D. Fontana, G. E. Kugel: Phys. Rev. B **41**, 12362 (1990)
4.120 P. C. M. Planken, L. D. Noordam, J. T. M. Kennis, A. Lagendijk: Phys. Rev. B **45**, 7106 (1992)
4.121 H. J. Bakker, S. Hunsche, H. Kurz: Phys. Rev. Lett. **69**, 2823 (1992)

4.122 K. P. Cheung, D. H. Auston: Phys. Rev. Lett. **55**, 2152 (1985)
4.123 H. J. Bakker, S. Hunsche, H. Kurz: Phys. Rev. B **50**, 914 (1994)
4.124 H. J. Bakker, S. Hunsche, H. Kurz: Rev. Mod. Phys. **70**, 523 (1998)
4.125 F. Vallée, C. Flytzanis: Phys. Rev. Lett. **74**, 3281 (1995)
4.126 S. B. Fleischer, B. Pevzner, D. J. Dougherty, H. J. Zeiger, G. Dresselhaus, M. S. Dresselhaus, E. P. Ippen, A. F. Hebard: Appl. Phys. Lett. **71**, 2734 (1997)
4.127 M. Hase, K. Mizoguchi, H. Harima, F. Miyamaru, S. Nakashima, R. Fukassawa, M. Tani, K. Sakai: J. Lumin. **76-77**, 68 (1998)
4.128 R. W. Schoenlein, D. M. Mittleman, J. J. Shiang, A. P. Alivisatos, C. V. Shank: Phys. Rev. Lett. **70**, 1014 (1993)
4.129 D. M. Mittleman, R. W. Schoenlein, J. J. Shiang, V. L. Colvin, S. P. Alivisatos, C. V. Shank: Phys. Rev. B **49**, 14435 (1994)
4.130 J. L. Machol, F. W. Wise, R. C. Patel, D. B. Tanner: Phys. Rev. B **48**, 2819 (1993)
4.131 U. Banin, G. Cerullo, A. A. Guezelian, C. J. Bardeen, A. P. Alivisatos, C. V. Shank: Phys. Rev. B **55**, 7059 (1997)
4.132 T. D. Krauss, F. W. Wise: Phys. Rev. Lett. **79**, 5102 (1997)
4.133 E. R. Thoen, G. Steinmeyer, P. Langlois, E. P. Ippen, G. E. Tudury, C. H. Brito-Cruz, L. C. Barbosa, C. L. Cesar: Appl. Phys. Lett. **73**, 2149 (1998)
4.134 R. E. Slusher, L. W. Hollberg, B. Yurke, J. C. Mertz, J. F. Valley: Phys. Rev. Lett. **55**, 2409 (1985)
4.135 M. Artoni, J. L. Birman: Phys. Rev. B **44**, 3736 (1991)
4.136 X. Hu, F. Nori: Phys. Rev. Lett. **76**, 2294 (1996)
4.137 G. A. Garrett, J. F. Whitaker, A. K. Sood, R. Merlin: Optics Express **1**, 385 (1997) http://epubs.osa.org/opticsexpress
4.138 M. U. Wehner, M. H. Ulm, D. S. Chemla, M. Wegener: Phys. Rev. Lett. **80**, 1992 (1998)

Index

Topics in Applied Physics Founded by Helmut K. V. Lotsch

1 **Dye Lasers** 3rd Ed. Editor: F. P. Schäfer

2 **Laser Spectroscopy** of Atoms and Molecules. Editor: H. Walther

3 **Numerical and Asymptotic Techniques in Electromagnetics** Editor: R. Mittra

4 **Interactions on Metal Surfaces** Editor: R. Gomer

5 **Mössbauer Spectroscopy** Editor: U. Gonser

6 **Picture Processing and Digital Filtering** 2nd Ed. Editor: T. S. Huang

7 **Integrated Optics** 2nd Ed. Editor: T. Tamir

8 **Light Scattering in Solids I** 2nd Ed. Editor: M. Cardona

9 **Laser Speckle and Related Phenomena** 2nd Ed. Editor: J. C. Dainty

10 **Transient Electromagnetic Fields** Editor: L. B. Felsen

11 **Digital Picture Analysis** Editor: A. Rosenfeld

12 **Turbulence** 2nd Ed. Editor: P. Bradshaw

13 **High-Resolution Laser Spectroscopy** Editor: K. Shimoda

14 **Laser Monitoring of the Atmosphere** Editor: E. D. Hinkley

15 **Radiationless Processes in Molecules and Condensed Phases**. Editor: F. K. Fong

16 **Nonlinear Infrared Generation** Editor: Y.-R. Shen

17 **Electroluminescence** Editor: J. I. Pankove

18 **Ultrashort Light Pulses** Picosecond Techniques and Applications 2nd Ed. Editor: S. L. Shapiro

19 **Optical and Infrared Detectors** 2nd Ed. Editor: R. J. Keyes

20 **Holographic Recording Materials** Editor: H. M. Smith

21 **Solid Electrolytes** Editor: S. Geller

22 **X-Ray Optics.** Applications to Solids Editor: H.-J. Queisser

23 **Optical Data Processing.** Applications Editor: D. Casasent

24 **Acoustic Surface Waves** Editor: A. A. Oliner

25 **Laser Beam Propagation in the Atmosphere** Editor: J. W. Strohbehn

26 **Photoemission in Solids I.** General Principles. Editors: M. Cardona and L. Ley

27 **Photoemission in Solids II** Case Studies Editors: L. Ley and M. Cardona

28 **Hydrogen in Metals I** Basic Properties Editors: G. Alefeld and J. Völkl

29 **Hydrogen in Metals II** Application-Oriented Properties Editors: G. Alefeld and J. Völkl

30 **Excimer Lasers** 2nd Ed. Editor: Ch. K. Rhodes

31 **Solar Energy Conversion.** Solid State Physics Aspects. Editor: B. O. Seraphin

32 **Image Reconstruction from Projections** Implementation and Applications Editor: G. T. Herman

33 **Electrets** 2nd Ed. Editor: G. M. Sessler

34 **Nonlinear Methods of Spectral Analysis** 2nd Ed. Editor: S. Haykin

35 **Uranium Enrichment** Editor: S. Villani

36 **Amorphous Semiconductors** 2nd Ed. Editor: M. H. Brodsky

37 **Thermally Stimulated Relaxation in Solids** Editor: P. Bräunlich

38 **Charge-Coupled Devices** Editor: D. F. Barbe

39 **Semiconductor Devices for Optical Communication** 2nd Ed. Editor: H. Kressel

40 **Display Devices** Editor: J. I. Pankove

41 **The Computer in Optical Research** Methods and Applications Editor: B. R. Frieden

42 **Two-Dimensional Digital Signal Processing I.** Linear Filters Editor: T. S. Huang

43 **Two-Dimensional Digital Signal Processing II.** Transforms and Median Filters. Editor: T. S. Huang

44 **Turbulent Reacting Flows** Editors: P. A. Libby and F. A. Williams

45 **Hydrodynamic Instabilities and the Transition to Turbulence** 2nd Ed. Editors: H. L. Swinney and J. P. Gollub

46 **Glassy Metals I** Editors: H.-J. Güntherodt and H. Beck

47 **Sputtering by Particle Bombardment I** Editor: R. Behrisch

48 **Optical Information Processing** Fundamentals. Editor: S. H. Lee

Printed in the United States
92169LV00003B/257/A

9 783540 660859